New Directions in Digital Poetry

INTERNATIONAL TEXTS IN CRITICAL MEDIA AESTHETICS

VOLUME #1

Founding Editor
Francisco J. Ricardo

Associate Editor
Jörgen Schaefer

Editorial Board
Rita Raley
John Cayley
George Fifield

New Directions in Digital Poetry

BY

C.T. FUNKHOUSER

continuum

The Continuum International Publishing Group
80 Maiden Lane, New York, NY 10038
The Tower Building, 11 York Road, London SE1 7NX

www.continuumbooks.com

www.newdirectionsindigitalpoetry.net

Library of Congress Cataloging-in-Publication Data
A catalog record for this book is available from the Library of Congress

ISBN: HB: 978–1–4411–6592–3
PB: 978–1–4411–1591–1

Typeset by Fakenham Prepress Solutions, Fakenham, Norfolk NR21 8NN
Printed and bound in the United States of America

Contents

Introduction

The present work is the inaugural text in this new book series, *International Texts in Critical Media Aesthetics*, of which, as founding editor, I should say something regarding the rationale for its existence, particularly in relation to the field of contemporary art and literature. This series exists to provide a platform for rigorous scholarship in the now-prominent rise of new media studies as relates to their specific relevance to the aesthetic and the poetic – architecture, the arts, and literature. In the last four decades, significant scholarship in electronic art and literature has emerged, as evidenced by an observable rise in the number of journals, books, conferences, and festivals devoted to this field. In the mid-1980s, computer science scholarship emerged for the study of the then-novel process of hypertext and its place in the creation of literature whose reading was interactive and combinatorial rather than strictly temporal or linear in narrative. The primary venue for this scholarship was the annual Hypertext conference of the Association for Computing Machinery. The adoption of literary and aesthetic work by computer scientists with interests in the literary and visual arts was a natural outgrowth of experiments carried out on the effect of on-screen organization of text and image on readership styles as well as on the persistent question of whether it was possible to program a computer or other electro-mechanical system to create new kinds of visual or literary experiences. Principally in the United States, Latin America, and Europe, scores of now-classic systems were designed from the 1980s to the 1990s for automatically creating literature, poetry, and visual art. These systems in turn generated thousands of works of literature and art that were rendered on computer monitors, projected on walls, printed on large-scale plotters, or realized as autonomous moving systems of sculptural performance. Hypertext and other approaches to electronic literature were almost completely ignored by scholars of traditional print. Regarding visual studies, too, some artists involved in this line of effort, including

Nam June Paik and Jean Tinguely became documented within the art historical canon – largely because they were in contact with art critics of their day – but most worked in obscurity. However, electronic art and literature remained fertile ground for creative experimentation and by the late 1990s, events like the first *Digital Arts and Culture* conference in Bergen, Norway made explicit the continuation from original trajectories that were first inaugurated much earlier with *Cybernetic Serendipity*, the landmark exhibition of computer art shown at the Institute of Contemporary Arts, London in 1968. From the early work of these four decades and since then, the creative directions currently followed by certain new media artists, critics, and authors, media studies scholars have produced important historical, conceptual, and critical documents that have begun to lay a foundation for a domain of study that works with, but productively extends, art history and literary theory.

However, precious little of scholarly interest in what we might call new media studies (a vague domain of study that has assumed many monikers) has rigorously focused on the essential in literature, and even less on visual and performing arts. Much writing has often looked at the *instrumentation*, particularly historical, arcane, and obsolete media, that creates literary or visual experience. That emphasis, which has been termed media-historical, is somehow distinct from the experience of the literary, poetic, or aesthetic. The analysis of experience, while not *wholly* independent of the media that deliver creative works, is nonetheless sufficiently autonomous that there remains something far beyond the physical or material support which, as in the past, stands in need of rigorous examination on its own terms. That is why the series of which the present book is the inaugural volume contains the indication of that focus – *critical media aesthetics*. The first word in that term reminds us that philosophy's notion of the *critical*, particularly in German and French efforts during the twentieth century, assumed a position of distance from observable phenomena, and from that vantage, could objectively question the many ideologies of instrumental ideology (for all instruments *are* ideologies). Hence, a critique of media aesthetics (rather than of media alone) implies that the creation and reception of word, image, and object are now complicit with new media but must be examined in a rational rather than an instrumental perspective. Were

John Dewey alive today, he most certainly would not be thinking about machines, but rather of the creative forces that obtain in any setting regardless of medium utilized.

Thus speculations about the relationship between creative impulse and the way that it is organized over a medium must come before discussions of the medium itself. That mode of organization, for example, informs the idea of *genre* in any creative practice, so that one of the ways in which visual art and poetry connect intimately is through play with the materials of the medium, but that play is first determined by artistic imagination, not by the medium. In the poetic work, one explores not merely a question of *meaning* but also an encounter that is a play with language, structured and conveyed for a new level of meditative realization. Much of what is fulfilling in both a visual work and in a poem is the sense that what is brought to us is interesting, while the materials, the medium, the context of support – these, too, are equal constituents of what could be called the poetic or aesthetic experience. This conjoining of medium with effect is not unique to works of art or poetry, nor is it a new experience, for as Manfredo Tafuri reminds us, the arcades and department stores of Paris, like the great expositions, were certainly the places in which the crowd, in addition to the architecture, became the spectacle.[1] As with the experience of traversing an arcade, meandering through the structures of digital poetry brings us to the realization that the environment is as fluid and crucial to the reflective experience as is the visible content itself.

This expedition, more necessary now, after the appearance of new kinds of dynamic web-based media, of installations, and of other kinds of 'arcades' through which the poetic work can be acquired, has made the present book indispensable as a guide and as a conceptual companion to the structure and interpretation of whole classes of poetic works in realized and deployed through and within electronic media. Sharing the manuscript with two colleagues versed in traditional print literature, I saw the same reaction – it was one of surprise at the uniqueness and variety of the relationship that a reader encounters in digital poetry with new structures and methods that evoke the poetic aura so familiar to anyone who has glimpsed the essence of a poem. For, as with his 2007 book *Prehistoric Digital Poetry: An Archaeology of Forms, 1959–1995*, and following its arc,

Funkhouser again brings us an array of close readings of electronic poems – comprising works created after the World Wide Web – that can be read as a layperson's guide to the process of experiencing digital poetry at present. In this approach, discussions of interface behavior interweave with questions of symbolic inference as emerges from the text's own content, and the unified experience makes evident that whole new modalities of reading are afforded the reader. Read after his 2007 book, the present work's case studies address the question, 'What happened to poetry after the birth of the Web?' And in the work of many poets of electronic form – John Cayley, Deena Larsen, Jim Rosenberg, Alan Sondheim, Jody Zellen, Serge Bouchardon, Jim Carpenter, Angela Ferraiolo, Mary Flanagan, Aya Karpinska and Daniel C. Howe, mIEKAL aND, Eugenio Tisselli, Eric Sérandour, Philippe Bootz, Talan Memmott – the medium has opened new creative terrain that is unprecedented in its creative potential, and Funkhouser's careful treatment of both poet and work is unusual in breadth and probing depth. Several typological differences in the structure of poetic work comprise the logical divisions of the book, most notably regarding works that rely on the World Wide Web for replication, although each copy operates as an autonomous and hermetic poem, versus those that co-opt the global network by harnessing its data streams, its ability for synonym generation, and its image and text retrieval capabilities, so that the engineer's notion of 'real-time' serves the poet's work as an additional aleatory dimension.

The emphasis on structure and organization of material process *over* that of the functionality of a medium is present not only in Funkhouser's approach in this book, but is also exemplified by the striking image on the front cover. The work in question, *belled*, is by the digital artist Henry Mandell, whose own approach is both typographic and visual. Using text as the raw visual ingredient in his work. Mandell transforms a written passage about the acoustic properties of temple bells by means of individual manipulations of the text's typographic characters. Since they exist as objects of vector-based typography, Mandell's works lie at the intractable intersection of the textual and imaginal. These properties – which are not device independent – allow the visual elements of the work to be stretched unremittingly until the constituent text loses its symbolic quality and presents instead an indexical image of the acoustic properties of its subject matter. Perhaps that chrysalis is the best descriptor of the aim of the series, and of the present book.

Francisco J. Ricardo
Cambridge, MA
September 2011

Things I saw on my drive up!
double rainbow
Swaguara
Farmers Daughter

1

Poetic mouldings on the Web

They are all laid out before us: the genuine post-modern text rejecting the objective paradigm of reality as the great 'either/or' and embracing, instead, the 'and/and/and'[1]

JANE YELLOWLEES DOUGLAS

The creative task of digital poetry often involves an artist observing and making connections between separate but poetically associable entities and then using technological apparatuses to communicate to an audience through compelling presentations. Jim Andrews' online digital poem *Arteroids*, for example, borrows its stylistic cues from a 1980s video game called *Asteroids*. For years, I have introduced students to this work, perplexing them with an assignment unlike anything they have experienced before: a video game featuring fragmented language proposed as poetry – or poetry designed as a type of game without competitive structure. Upon study they begin to understand how digital poetry functions as something other than poetry presented on a computer, involving processes beyond those used by print-based writers, and that poetry made with computers has unusual qualities – representing something inventive and worthy of engagement.

Students in the Senior Seminar on Literature, many of whom are familiar with the original game, appreciate Andrews' work from a variety of viewpoints.[2] Their responses reflect how they access literature through novel perspectives located well within the domain of contemporary culture, and enjoy and learn through the provocation of receiving crafted language amidst non-traditional methods. The engagement becomes exciting due to atypical demands foisted upon them as they explore and discover ways to understand the experience.

One recent student in the course clearly understood distinct poetic attributes of *Arteroids* and identified the artistic function presented through its language: 'I noticed two levels of meaning in the words: the first was that the words themselves represented the context of the player and/or what the player was doing or feeling. ...On a second level, the words when put together in random context made phrases that not only made sense but related to the poem: 'Destroy poetry', 'create poetry', 'make me what I am into another', were phrases relating to what I was doing as a player to the 'arteroids/words'.[3] In this regard, Andrews' work (as well as other examples of digital poetry) may teach its viewers something similar to what Dada did: that a string of phrases appearing to be nonsensical on the surface in fact contains meaningful poetic logic. An accumulation of fragments, taken as a whole, can assert profound recognitions. Another student working to cultivate a literary background observed the work from an alternative perspective, receiving a different type of learning lesson:

> I am slowly growing to develop a substantial appreciation for poetry and 'dullness' is one of the things I tend to struggle with. I was indeed battling different colour variations of the word 'Poetry', which I found to be symbolic, like I was actually carrying out the goal of making poetry suffer; physically the game started to get more stressful as I was surrounded by words that were related to prior forms of poetry and it was almost like an assault. As I continued to play I became less involved with the words and more engrossed with destruction and survival.[4]

Here the student looks for and finds something exciting within *Arteroids*, despite distractions presented. She begins to see and sense poetry as a concept requiring action, and experiencing Andrews' piece engages her; the activity and concept of poetry becomes personified through the game, and serves as a reinvention of the form.

Comments such as these show how a digital poem can have multiple personalities, and that different viewers see and attend to dissimilar attributes emanating from within the same work. Digital poems can be self-reflexive with regard to what the reader does and what the poem does, and also ignite other types of cognitive activity through its multi-sensory layering. Most students find ways to value the vivid experience of *Arteroids* – that they do more than just 'stare at the computer' and that they can, given Andrews' design, create their own custom word base with which to populate the game. Approaching composition interactively, in the form of a simple game with a quirky soundtrack and variable database, impresses them because it completely changes the act of reading (or writing) a poem and creates an ambiance previously unknown to poetry. Unique settings and multimodal expression for artworks proliferate as digital technology and networks develop.

Foreground

Noah Wardrip-Fruin defines digital literature as 'a term for work with important literary aspects that requires the use of digital computation'.[5] As such, it takes on many forms (as works presented on PCs or mobile devices, as installations, as performances), and presently exists as a creative possibility for artists around the world. Communities of different types of digital writers have blossomed. One of its major subsets, digital poetry, to its credit, cannot be singularly defined. Authors of digital poetry engineer artworks featuring a wide range of approaches and styles (forms) of construction rarely presented as straightforwardly as poems composed for the page. Digital poetry can be seen as a type of organism, as an approach to expression having properties and functions determined not only by the properties and relations of its individual parts, but by the character of the whole parts, and by relations of parts to the whole. Thus far, in its relatively brief history, an ever-present variability has proven itself as a primary attribute of composition (and thus presentation). Digital poetry, as a literary and artistic form, is an equivocal organism, with many identities or iterations. As an expressive form, it matters not only as a free-ranging serious practice, but because it invites vibrant, transformative multimodal engagement for its practitioners and audience alike.

This book gives an account of processes involved with experiencing and understanding digital poetry, a genre complicated by the variety of forms it embraces. My analysis focuses on the World Wide Web (WWW), at present the public's predominant point of intersection with digital poetry, where audiences privately absorb and/or interact with texts. The WWW offers an unprecedented venue for the circulation of digital poetry, some of which depends on the network in order to function. Writers and artists create with computers in ways that do not simply document the poetic forms of bygone eras: they are reinventing the possibilities for poetry. Computer processes, still relatively new to the world and artists engaging with them, adorn poetic features unavailable to previous generations of literary artists. The intimate, yet often impersonal, location of the network often demands negotiation of an exorbitant amount of information, which tends to obfuscate the encounter.

Digital poetry may faintly – and not so faintly – remind us of various types of historical approaches to writing or verbal articulation. At times we see contemporary enactments of previously established approaches to vibrant artistic expression drawn from multiple forms (e.g. multimedia collaborations at Black Mountain, or Fluxus), although these mediated combinations rarely yield a clear-cut presentation of materials. Many titles do not reify poetic forbears; highly calculated, complex digital poems on the WWW achieve previously unattainable literary effects.

Digital poetry appeals to me because it offers forms of artistry inviting (and uniting) processed interconnections between sound, image and language. Further, while variably inscribing each of these components, digital poems avoid simple iterations of them – permitting their use, often accentuated with degrees of randomness, fracture, and even poetic disconnection between media. Literary work, on the network, excitingly transpires on multiple registers: on thematic levels, by using artful language, and through responsive technical achievement. Advancements resulting from technology innovate to profound degrees: practices discussed below intentionally present poetic alternatives to the WWW's general ontology of promoting products and serving up data at rapid speeds. These endeavours result from cultural predicament, as a foil to operative conditions now predominant on the network.

Interconnection insinuates a whole made of parts, or hybridity. Hybrid approaches to the construction of language, with or without media, requires use of fragments, as in choosing letters and phonemes to make words. Digital literature, generally speaking, exhibits fragmentary authorship – its penchant for fracture and decentralization are proven, and multimedia poems partake of this tradition. Considering this subject, we must acknowledge that by definition this implies something broken into pieces, whose parts may never fully reform. Artworks employing atomic (or, culturally speaking, post-atomic) techniques engage with such conditions, often repudiating complete sensibility. Even the most heavily disjointed works, however, provide ways for readers to make stimulating connections. If a work's contents are not, and cannot be, fixed, we can, even if momentarily, 'fix' or build potentially profound understandings of these works on an individual basis (if not within larger, all encompassing categories); patient, observant perusals offer rewards.

While synthetic in essence, and brittle in terms of longevity, digital poetry's fluid states prevent us from considering works as being plastic. Yet because they never harden, works of digital poetry always maintain *plasticity* in presentation on the WWW. They exist in a state of being moulded, receiving shape, made to assume many forms – often seeking qualities that depict space and form so as to appear multi-dimensionally. When biological organisms with the same genotypes have the capacity to vary in their developmental pattern, in their phenotypes, or in their behaviour according to varying environmental conditions, scientists say they have plasticity. While the organic component of digital poetry primarily involves human input in production and consumption, recognizing plasticity as an aesthetic foundation establishes a valuable metaphor for generally qualifying the results of electronic writing to date. A given, though gradually increasing, number of formal foundations exists for authors to work with, yet titles abundantly vary due to technological

I miss them when they're gone."

flexibility. We have, in digital poetry, both aesthetic dependability (overt foundations) and *deep bendability*.

Attaching significance to the medical connotations of this metaphorical aesthetic sensibility sheds further light, I believe, on what digital poets add to the parameters of literature. Plasticity refers to the capacity for continuous alteration of the neural pathways and synapses of the living brain and nervous system in response to experience that involves formation of new pathways and synapses and/or modification (or elimination) of existing ones. As readers of digital poetry, we must ourselves become mouldable, capable of reshaping ourselves and our expectations on a text as a whole depending on what we encounter on the screen. These artworks require reformed conceptions, extending parameters and dimensions of reception beyond those we ordinarily use to absorb or experience expression.

The great extent to which authors choose to render plasticity complicates processes of reading. Innovative poetry has always challenged readers who seek straightforwardness, but use of media compounds the intricacies of expression. Beyond making literally difficult-to-read works via material obstruction or speed of presentation, authors commonly implement unconventional, disconnected syntax and phraseology; verbal elements found in these works rarely present ordinarily arranged language. Engaging with digital poetry requires more from readers, who face multimodal, human-to-machine transcreations where texts initially presented in one state transform into others. These assets assist in elevating contemporary titles to poetic realms, but decentre the text proper's authority. As poetry becomes a networked form, its poetics explodes and singular measurements of its pliancy resist finite definition. Digital writing, on and off the WWW, does in part borrow processes and combine methods from the past. What emerges sometimes corresponds with efforts by artists whose unmediated inclinations resisted convention through openness to form and abstraction, with vibrancy and difficulty now added due to technological foundations and the breadth of range in these new practices.

My 2007 book *Prehistoric Digital Poetry: An Archaeology of Forms, 1959–1995* provides a detailed account of how artists (and programmers) used computers to foster the new genre of digital poetry,[6] illustrating how computers gave poets new modalities with which to present language-based expression in various types of activated, visual, interconnected forms before the WWW existed. At the time, I felt it was important to engage with research in this area because descriptions (and understandings) of the types of poems appearing on the WWW rarely included fair references to the works which preceded them in the literary continuum. *Prehistoric Digital Poetry* illustrates how most of the mechanical groundwork for digital poetry was already established by the time the WWW became prominent. The genre's fundamental building blocks and textual parameters – cohering as generated, visual and hyper texts – were essentially solidified by the time the first webpage was ever made. With its versatility and capability to bring the disparate forms of expression together, the WWW marks the place where expressive digital forms synthesize, become more widely circulated, and with the increase of demographics and technological advancements, a flexible genre – partially associated with its written counterpart – forms. At the juncture when the WWW appears, the prosody of digital poetry already includes text generation, flexible and collaborative language, interactivity, intertextuality, and applications of various visual and sonic attributes. More than a decade later, different conditions for presentation exist, yet principal elements of the genre remain central, even if progress and innovation occur by way of works drawing directly (i.e. instantaneously) from the network for content.[7] In terms of historical trajectory, recognizing that most digital poems appearing on the WWW in its first decade did not specifically rely on the network is important; these works do not depend on the WWW, but benefit from it as a publishing platform and the manner in which the WWW's attributes enable presentation of multimodal works. Instead of being formally distinct, the building blocks of composition begin inextricably (if variably) to merge with one another within a single title. The genre has not so much expanded as it has evolved to fuse its disparate characteristics and perpetually invents alternative approaches by which to stage and facilitate its plasticity.

For accessibility and focus, this book concentrates on publicly accessible work, made with contemporary technologies and circulated on the WWW. In contrast to *Prehistoric Digital Poetry*, the majority of subject matter remains available; anyone with an Internet connection can view these digital poems. That said, a few programs I studied and wrote about while composing this book have disappeared, and may be gone from public view forever – a significant factor of digital textuality. Eventual hardware, software and network modifications (e.g. 'upgrades' that squelch capabilities of previous versions of programs) make inevitable the extinction of certain, if not entire swaths of, works. Unfortunately such departures can happen quickly and unexpectedly. For example, two works profiled in the case studies below, *Erika* and *Google Poem Generator* – emblematic of types of poems produced during the WWW's first decade – no longer appear on the network.

The present study partially surveys the initial phases of digital poetry on the WWW. Hundreds of worthy digital poems have been authored, of which I discuss a few, covering a fraction of what has been produced. To write a complete history of digital poetry since the WWW became its primary textual staging area, given the proliferation of creative expression and development of progressive formations on the network, would proceed endlessly. A comprehensive survey analyzing every significant title would entail more pages than a book could hold. Instead, I take up a new objective by directly, pragmatically addressing the matter of, and processes involved with, viewing and grasping the varied and sometimes interactive works gathered under the rubric of digital poetry.

Given its plasticity, one of electronic literature's foremost challenges, understanding poems designed for presentation on computer networks, requires guidance. While preparing this book, I gave a talk at the Maryland Institute for Technology in the Humanities (MITH) titled 'Digital Poetry as Scrabble: Making from Given Materials', during which I showed examples of digital permutation poems. The audience included several literature professors whose research focuses on conventional poetry written long before the digital era. They became curious, even fascinated, by the material, but confessed they had no idea how to read what was shown.[8] These readers, uninitiated in the process of approaching and comprehending a digital poem, needed a guide. Beyond delivering contexts for the artistic practices involved, this book discusses of a variety of titles indicating what readers may expect while perusing digital poems on the WWW, and how to handle observing and participating in such encounters. Since the WWW has emerged, dynamism and intricacy of presentation of works produced in the genre have increased. These developments complicate the reception of literary expression due to their lack of general uniformity, even if the majority of titles predominantly feature similar foundations as works pioneered before the WWW era. Aesthetically speaking, as acknowledged in Jane Yellowlees Douglas' 1991 observation about encountering hypertexts (used as epigraph above), viewers of these poems do not confront texts containing one attribute or another, but rather texts inscribing multiple attributes combined together. Points of identification in digital poems always shift and have capacity to contain many variable attributes. In Douglas' words, they embrace the 'and/and/and', an effect that characteristically mystifies their reception.

WWW as mechanism

Because technological apparatuses used by digital poets directly influence what they produce, I will now briefly trace the development of some of the tools that significantly enable works discussed in this book. Mediated capabilities inherent in the WWW have gradually expanded since its initiation in 1991 (as an extension of the Internet), endowing it with resilience and artistic potential. Initially proposed as a 'large hypertext database with typed links', the WWW's purpose was 'to allow high energy physicists to share data, news, and documentation'.[9] Hypertext, a concept and non-linear textual practice first introduced by Ted Nelson in the 1960s, envisioned writers and designers creating 'branching' structures in literary documents (including video), connected by links embedded in texts and imagery presented. Early iterations of the WWW, however, prohibited the fluid transmission of such media, and a complete realization of this objective did not begin to occur until 1995, when HTML code[10], readable by WWW browsers, began to enable embedded scripting codes (e.g. JavaScript), thus expanding expressive capabilities of the network by supporting integrated multimedia in addition to hypertext linking capabilities.[11] Two other developments elevated the network's capabilities, increasing its versatility for artists: in 1999, the bandwidth became 2.5 Gbps (Gigabits per second);[12] a year later, refinement of Internet Protocols responsible for the delivery of 'packets' occurred, which ensures the correct delivery of data.[13] As the Internet expanded and many more computers came online, these developments crucially enhanced the functionality of the network and made an impact on the types of materials artists could produce with confidence (although it should be reiterated that most works will function offline and do not actually depend on the WWW for their presentation). Increased bandwidth meant files containing larger amounts of information, as in videographic material, could be more rapidly transmitted and received across the WWW. For example, QuickTime movies, produced using software technology made available in 1991, became a more feasible way to present art across the network.[14] During the 1990s, an array of video software programs (and players) influenced the conception and presentation of digital poems.

In the WWW's early years, cultivation of software and integrated computer programming, along with advancements in network performance, significantly altered the organization and rendering of the new modalities of literary expression.[15] With the release of the software program Macromedia Flash 3 (1998), the ability to produce animations and interactive documents through simple scripting commands became much improved.[16] As of 1999, MPEG-1 (mp3) technology enabled production of audio files in a highly compressed format conducive to computer playback. Shortly thereafter, Flash 4, which supported the use of mp3 files as well as the 'Motion Tween' function, diminished labour in authoring animations. The importance of Flash – a program with which works are composed on a 'stage', using a timeline – on the manufacture of digital poetry in the early WWW period cannot be understated. Because, as Nelson observes, Flash is 'essentially a complete platform – a complete program package with all forms of graphics, interaction and communication', multimedia poems of all sorts have blossomed.[17] Notably, Flash software, written in C++, evolved to include an object-oriented programming language (ActionScript 3.0). Thus the program itself is a hybrid compilation (or formation) of programming codes.[18]

Many new technologies developed since the mid-1990s contribute to innovation and make an impact on the aesthetic conditions of digital poetry.[19] For example, digital poets began using cascading stylesheets (CSS, see *Stir Fry Texts*, Ch. 3), which, in addition to other tasks, facilitates the uniform appearance of pages in a complex work. Scripting interfaces such as ASP (Active Server Pages, developed by Microsoft) effectively created unusual automatically generated poems across the network, as in Leevi Lehto's *Google Poem Generator* (Ch. 5). Numerous artists have used Java, an object-oriented programming language, to build interactive and animated poems with dynamic content on the WWW.[20] PHP (1995), another object-oriented scripting language, acts as a filter taking input from a file (or stream containing text) and outputs another stream of data in HTML; such processes importantly contribute to the cultivation of dynamic web pages, as in Eugenio Tisselli's work (Ch. 5). Lesser-used but effective methods and programs, some of which involve object-oriented paradigms, also materialized (e.g. Squeak, used in Jim Rosenberg's *Diagram Poems*, Ch. 3).[21]

Altered conditions for textuality and a pluralized, unfixed aesthetics result from these advancements. Mechanisms permitted by the WWW literally and figuratively heightened the visibility of digital poetry on the Internet.[22] With every new browser release, the WWW became more versatile, easier to use, and more interesting. Sites were static at first, e.g. Diana Reed Slattery's 1996 title *AlphaWeb*, then animations and artworks containing sound – multimedia art and poetry – began to appear with frequency e.g. Christy Sheffield Sanford's 'Flowerfall' or Mark Amerika's *Grammatron*, both produced in 1997.[23] For media-inclined poets, the WWW's technological growth brought expressive opportunity and variety, providing continuum and stylistic advances. Multiple surfaces and components layer in the poetry – already a complex expressive form – a compound of written/visual/audio/mental expression with code, software and the network, fluidly enabling media and intertextual potential. In this book, we observe the impact of these factors on poetic forms becoming evermore refined, interactive and composite in the WWW era.

Fused Forms and Types

Increased levels of interactivity, and the ongoing elimination of barriers between digital forms of writing (generated, visual, hyper texts), reposition digital poetry's textual conditions. WWW technologies enable modalities combining text, graphics, animation/video, sound, interactivity, generative properties, randomness and permutation. Identifying singular aesthetic considerations upon which to define forms and types of digital poetry is now a burdensome, if not impossible, task. In this sense, digital poetry is unquestionably 'noisy', to borrow a construct from Brian Kim Stefans' 2003 title *Fashionable Noise: On Digital Poetics*. Distinctions between forms, incisively established for somewhat more primitive (but not always quiet) works in *Prehistoric Digital Poetry*, remain useful as a way to identify individual components or methods used by digital poets, but can no longer be used to divide them categorically. In many examples discussed below, forms that had been previously separated now fuse together, enhancing and boosting the density of works produced in the genre.

Prehistoric Digital Poetry establishes three major forms of artistic production, as well as a few fringe areas. The first digital poems, then sometimes called 'computer poems', came from programs generating poems by uniting a database and a series of commands.[24] While this fundamental practice (using computer code to process language) has continued on the WWW in works such as Jim Carpenter's *Erika*, Millie Niss' *The Electronic Muse* and many other titles, core materials now capably incorporate graphics and other multimedia effects. Visual and videographic poems exhibit lively generative features, as in Angela Ferraiolo's *The End of Capitalism*, and generative texts surgically draw from extensive online resources (such as Google, in the case of Lehto's *Google Poem Generator*). Poets not only write programs to compose poetry but programs that automatically synthesize alternative media elements into digital poems, as in Tisselli's *Synonymovie* or Andrews' *dbCinema*.

Visual digital poems and videopoems started emerging in the 1960s, successfully literalizing the cliché 'poetry in motion'. In some regards, contemporary practice follows historical models, but since the demographics have broadened, deviation from the principles of concrete poetry and other historical predecessors has occurred. Visual poets such as K. S. Ernst, Geof Huth and others, continue to create digital works, and most digital poems contain noteworthy visual aspects. Combinations of text, image, movement, sound and myriad types of stylistic shaping are imagined and produced. Many members of the extensive international community of visual poets use computers to produce graphical alphabetic works. Blogs, websites and listservers, such as Huth's *dbqp: visualizing poetics*,[25] *Crg Hill's poetry scorecard*,[26] and mIEKAL aND's *Spidertangle*,[27] are respectively among the many resources serving as outlets for dissemination of information and archiving works in this realm. Visual poetry is, in my view, a particular branch of digital poetry – but also exists independently, as part of the continuum of the great and varied lineage established by analogue visual poets. Though many artists work with computers, not everyone assigns significance to the fact that images made using Photoshop involve automated digital programming even if artists employ it to enact manual processes and manipulations. Digitally processed alphabetic information substantiates these works, and a finite study of the influence of computers on visual poetry should be written. Digital poetry inarguably relies on visual elements (through static, animated and videographic imagery, or graphical and visual treatment of language); graphical components are often central, even though works vary widely in terms of dynamics (note the difference between aND's *Seedsigns for Philadelpho* and Jason Nelson's *I made this. you play this. we are enemies*), and may feature non-linear or generative elements.

Numerous factors prevent establishing definitive, fundamental distinctions between poetic forms embracing digital media. Beyond 'the variety of approaches to digital poetry', writes Memmott in the essay 'Beyond Taxonomy: Digital Poetics and the Problem of Reading', 'the transitive aspects of its elements, and the transactive quality of its applications, make the development of a consistent, stabilizing taxonomy difficult if not impossible.'[28] In contrast to the pre-WWW era, very few literary hypertexts (poetry or fiction) present only text; even examples privileging text usually contain graphical design components. We might thus be tempted to regard all digital poetry as hypermedia, but cannot ignore that the hyper- prefix by definition assigns a non-linear dynamic to the work. Many works on the WWW, despite rich uses of media, perform linearly, even when authors produce sophisticated interactive structures within their visually-based verbal constructs, as in Serge Bouchardon's *Loss of Grasp* (Ch. 4).

Authors/programmers assemble works using two primary presentational strategies: projected and participatory. Projected digital poems do not entail interactive involvement; participatory poems require a reader's contribution to the production of narrative, and sometimes enable viewers to add new content within the participatory structure. Borrowing Michael Joyce's categorizations for hypertext/hypermedia forms in *Prehistoric Digital Poetry*, I identified two distinct approaches describing readers' roles, the 'Explorative' and the 'Constructive.'[29] Shifting the meta-framework for the now largely integrated, hypermediated WWW-based works to 'projected' and 'participatory' represents a re-positioning of textuality that falls in line with Espen Aarseth's senses of 'ergodic' and 'nonergodic' works respectively.[30] Just as digital poetry has become more dynamic, and in certain senses has made poetry more contemporary and dynamic by moving away from a fixed state, ways in which works can be theoretically and aesthetically built and categorized are multiple and fluid.

Digital poetry's plasticity invites inventive critical perspectives. For example, Stephanie Strickland introduces imaginative frameworks through which to understand contemporary practices in several essays, including 'Writing the Virtual: Eleven Dimensions of E-Poetry.' In this piece, Strickland outlines eleven 'entangled' conditions of 'Poetic e-writing', rather than charting poetics through more narrowly defined formal attributes.[31] Strickland's loose and even abstract contexts (e.g. '3. Time, become active, stratigraphic, and topologic, is written multiply') contain value as concise, inventive, knowledgeable views that remain open to interpretation.[32] 'Writing native to the electronic environment', observes Strickland, 'is under continual construction'; in this statement she acknowledges the transitory/transitional condition and 'instability' of the craft.[33] In order to cultivate a system open to invention, critics and practitioners benefit by not opting to evaluate digital poetry on technical or theoretical bases, or by speaking in absolutes. As digital poetry's

contexts). They explain individual writing strategies, authorial concerns, and the significance of our participation in such texts, illuminating digital poetry's adaptability and how varying senses of sequential events impose new thresholds of difficulty (and delicacy) to readers unacquainted with mediated expression.

The large range of critical approaches taken thus far confirms aesthetic dimensions and the overall character of digital poetic forms as being far from fixed or undemanding. Every element of digital poetry made for presentation on a dynamic network declares fair premises on which to discuss the topic, and variables within these forms absolutely complicate reception of the work, proven by the number of ways critics have invented to discuss them. Struggling to understand intentions behind a specific conglomeration of textual elements or explaining the reasons for an author's material choices inspires plentiful perspectives. Beyond matters of pure content and issues raised by traits such as indeterminacy, e.g. made-at-the-moment variable pathways, multiple endings, many factors amplify interpretative opportunities given the prevalent hybridity of forms.

Collecting Electronic Literature

In the opening chapter of *Electronic Literature*, N. Katherine Hayles writes, 'Will the dissemination mechanisms of the Internet and the Web, by opening publication to everyone, result in a flood of worthless drivel?'[37] In order to subvert the possibility, she co-edited the *Electronic Literature Collection* (ELC), Vol. 1, stamping approval on a wide range of digital texts worthy of preservation.[38] Many examples of digital poetry circulate in this quasi- canonical compendium. The editors – Hayles, Nick Montfort, Scott Rettberg, and Strickland – compile the first major institutionally funded collection of electronic literature, a historical event inasmuch as it provides crucial documen-tation of text representing the initial WWW era, artworks that Hayles asserts 'test the boundaries of the literary and challenge us to rethink our assumptions of what literature can do and be'.[39]

Expanding these boundaries in their own terms, the editors identify more than fifty general categories (style of work, software used or other demographic) to use as keywords describing the contents. Their signifiers classify apparent traits, establishing a clear vocabulary of terms enumerating both creative potential and worthy critical concerns: Ambient, Animation/Kinetic, Appropriated texts, Audio, authors from outside North America, Cave, Chatterbot/conversational character, Children's Literature, Collaboration, combinatorial, conceptual, constraint-based, procedural, critical/political/philosophical, database, documentary, essay/creative non-fiction, fiction, flash, games, generative, hacktivist, html/dhtml, hypertext, inform, installation, interactive fiction, java, javascript, locative, memoir, multilingual or non-english, music, network forms, non-interactive, parody/satire, performance/performative, place, poetry, processing, quicktime, shockwave, squeak, Storyspace, stretchtext, TADS, textual instrument, text movie, 3D, time-based, translation, viral, visual poetry or narrative, vrml, women authors, wordtoy (ELC).

This anthology represents the major historical forms of digital writing, in many synthesized forms. Digital literature has for more than three decades resisted, as if by definition, the need to embody a singular set of mannerisms, and this publication verifies the wide span of identifiable attributes and approaches used to produce works. Such editorial openness – integrating a diverse range of compositions – provides a true representation of the aesthetic identity of the genre. Looking closely at these works reveals how strategies towards making digital poetry has overtly evolved and diversified during the past two decades. For instance, readers familiar with digital poetry's past will discern that text generators are now easier to use, and often more versatile and participatory than their historical predecessors; a user no longer simply presses a button and watches the generated poem appear as ASCII text. Visual works have progressed beyond a state of static rendering; hypertext performs more viscerally than in the past. The anthology contains historical works both participatory and projected, such as Jim Rosenberg's *Diagrams Series* poems and Stefans' *the dreamlife of letters*. Rosenberg's online extension of his ground-breaking HyperCard work uniquely projects multilayered expression in which readers cull verbal content while negotiating visual information using syntactical formations invented by the author (see Ch. 3). Stefans' visually spectacular animated concrete poems provide a stunning example of how elastic and expressive letters themselves can be when an author employs new media techniques with great sophistication. Literature conceptually unites with computer games (e.g. Andrews, Ch. 3). A high level of interactivity employed by authors in the collection indicates that a greater number of works aspire to position viewers as players on a field rather than spectators in the grandstand – a disconcerting (yet potentially captivating) condition for those untrained to negotiate participatory works.

As a critic investigating this discipline, looking to identify its multifaceted characteristics, contributions to the anthology by early practitioners of the genre (e.g. John Cayley, Deena Larsen and Alan Sondheim) interest me not only for their plentiful artistic qualities, but also in terms of how authors who produced electronic work in the pre-WWW era have (and have not) shifted their compositional strategies, and, again, how the mediated properties of forms in general change over time. Hypertext, for instance, becomes more dynamic (although in certain regards possibly less sophisticated) than it was offline. Looking at Larsen's *Carving in Possibilities*, anyone familiar with her title *Marble Springs* (Eastgate, 1993) cannot help but notice how drastically hypertext techniques within a digital poem have changed. Instead of building links and nodes, presented and mapped separately, navigated by clicking with the mouse (as in *Marble Springs*), viewers now confront all the links at once; the process involves no clicking – moving the mouse across a given image unveils content. Several pieces recall historical works, such as Emily Short's *Galatea*, in which a 'chatterbot' emulates a conversation with the user (à la *Eliza* or MOO 'bot'), or the layered rays of text in Dan Waber and Jason Pimble's 'I You We,' which resemble the pre-WWW aesthetics of Marc Adrian's 'Computer Texts' or Carl Fernbach-Flarsheim's *The Boolean Image/Conceptual Typewriter*. While contemporary authors may not know these earlier projects, digital literature remediates their techniques in these recent titles, its fundamental formal building blocks perpetuated and advanced by a different cast of authors using a distinctive set of tools. Works gathered by the editors of this collection confirm the congruities and complexities of digital literature, showing how tremendous variety predominates in conglomerations of multimedia elements.

Polyformally Advancing

In a 1975 article titled 'Computer Art,' an early practitioner of digital poetry wrote: 'We need to find those things which uniquely suit these new media, which can only be expressed with their help, and thus make the effort worthwhile. I look for the fresh wind of ideas from the new wave of art students who will be literate in the information sciences, and conversant with interactive computers and the new processes which they can help visually explicate.'[40] Surveying the background of the digital poets whose mediated works I examine below, we find undisputable pluralism. Few have formally (institutionally) studied art. Works reviewed in my case studies were composed by individuals with backgrounds in mathematics (Andrews, Rosenberg), information science and communication (Bouchardon), education and business (Carpenter), chemistry (Ferraiolo), Chinese language and civilization (Cayley), social science (Mez), computer engineering (Tisselli) and English or creative writing (Larsen, Sondheim, Baldwin, Nelson, Strickland). Zellen, Mez and Flanagan studied art, but holding a degree in art is clearly not a prerequisite to becoming a digital poet. What has happened instead largely reverses Mezei's speculation: writers, mathematicians, and scientists and engineers of various sorts have become literate in programming and digital media and art.

These integrated artistic approaches, from wherever they arise, present new demands and expectations on viewers of digital poetry. Studying particular qualities and quantities contained in projected and participatory digital poems, while weaving a narrative about the process of understanding the poems, is my task. Building my investigation directly from groundwork established in my previous study might be sensible, but I only partly use *Prehistoric Digital Poetry*'s groundwork as a foundation due to omnipresent fusion of forms. Instead of using formal attributes as modes of demarcation, while respecting the value of celebrating continuum in the genre, I bridge the historical and contemporary by devoting the first of three analytical chapters to a discussion of digital poems presented on the network by authors whose practice begins in the 'prehistoric' era and now extends to the WWW. How do pioneers, those who worked in the field at an early stage, expand or alter their work via the network? As textual multiplicity reaches potentiality – as opposed to more strictly delineated formulations – how do the authors employ contemporary tools? A second analytical chapter explores titles by artists garnering attention subsequent to the WWW's emergence. Are these works less constricted than those made by artists who initially composed with fewer available possibilities? How, if at all, do they differ in orientation from those produced by artists of a previous generation? Works examined in both chapters extend historical forms, some forging unique inventive presentations, which are in turn advanced even further by works reliant on the WWW's mechanical properties (Ch. 5).

The field of digital poetry exists in a non-commercial realm. One might attempt to evaluate scientifically evaluate and rank practitioners based on the number of works they have created, or by how many performances or installations they have been invited to prepare or publish, but such equations would be problematic for many reasons. Many accomplished works have been produced, but only a few bring much public attention to the field. In the concluding chapter of *Prehistoric Digital Poetry*, I expressed a belief that no masterpieces had been created in the early years of the genre. The point is debatable, but the visibility of the practice in general is still limited due to various factors, not the least of which are difficulties brought on by the plasticity inherent in most works. Many artists have produced fantastic titles, some have been rewarded with jobs and paychecks as a result of their accomplishments, and a few artists receive exposure through exhibitions in major museums and galleries. This means an audience exists, and works are valued by those who are able to value them, but also that digital poetry is still nascent and maturing as a pluralistic artform. Artworks examined in this book represent important efforts produced by accomplished figures (although not the only practitioners); however, I sense the challenges they present may be prohibitive unless an audience is prepared to embrace them.

Works of digital poetry use computer technologies automatically or interactively to animate, shift (between content), shuffle and generate poems – interconnecting disparate textual or mediated elements persist at the nucleus of every digital work. Because materials vary from piece to piece, one cannot anticipate what to expect from a digital poem. In some cases text transforms into text, images transform into new images as text(s) transform from one state to another. We must remember the absolute rule of electronic text – the elemental key that puts it at odds with printed materials, a discrepancy so perfectly stated by Joyce's axiom, 'Print stays itself; electronic text replaces itself.'[41] Every example of digital poetry, by definition, steadily imparts content alternative to what initially appears; each screen stages different information, perhaps uniquely – and every change, each mediated shift, holds potential conse-quence. These conditions require keeping our attention focused on many disciplinary variables. Contemporary readers should not be fearful of embracing something new, or looking at possibilities for literature from all angles, and should enjoy the process. Plasticity and difficult consequences brought on by digital poetry and the superabundance of possibilities inherent in the genre need not lead to frustration; poetical celebration with exuberance, excess and surprise, conducted through media dynamics, has the capability to enthral once the organic functionality of the work is identified and understood.

2

Encounters with a digital poem

Reading remains a practice that is not reducible to information or to digital data[1]

JOHN LAVAGNINO (2007)

Context(s) for Reading

Readers of Shakespeare would concur that difficulties in reading poetry arose long before sophisticated technical engineering began to influence literary works. Carlos Drummond de Andrade's poem 'Looking for Poetry' (2002) reminds us how preparation plays a crucial role in any encounter with poetry:

Come close and consider the words.
With a plain face hiding thousands of other faces
and with no interest in your response,
whether weak or strong,
each word asks:
Did you bring the key?[2]

Digital poetry's significant dematerialization and ductility and the granularity of its treatment of language (atomic abstraction, fleeting dispensation), only further increase demands placed on its audience. Artists engaging in technological processes compound poetry's already complex foundations, resulting in the need for viewers to be prepared to approach the task of reading in new ways.

John Cayley's essay 'Writing on Complex Surfaces' (2005) explains why all forms of writing present challenges: 'The surface of writing is and always has been complex. It is a liminal symbolically interpreted membrane, a fractal coast- or borderline, a chaotic and complex structure with depth and history.'[3] As chronicled in *Prehistoric Digital Poetry*, a relationship exists between digital poetry and its Modern and Postmodern precursors. Acts of *reading* contemporary pliant forms, however, require a more participatory audience; as Bruce Andrews writes in 'Electronic Poetics', the reader becomes, in digital circumstances, 'the (modifying, reconfiguring) playback device, not the target of it'.[4] As such, readers must absorb, edit (or compile), and cognitively retain unconventionally presented data in order to approach comprehension.

Digital poetry presents difficulties, and exists at the fringes of literary arts precisely because of demands it imposes on audiences. Stylistically, modes of expression found in conventional and machine modulated writing differ greatly – a schism observed by Jed Rasula, who writes, 'In a curious way, Internet poetry can be compared to the American colonies circa 1760: full of enterprise and ingenuity, but a world apart, a realm of negligible consequence to an unheeding mother country and its culture.'[5] Poetry, for most, already represents an artistic aberration – something moving away from conventional diction and presenting perceptual challenges; digital poetry's inflation of this, through mediated expansion, may very well create too large a separation from norms and thus impedes garnering a larger readership. Similarities to typical conventions may exist in a digital poem – the output from a text generator often resembles a poem on a page – although many types of *reading* within the genre simply do not compare. Language and other pertinent content presented in multiple visual layers and sublayers within the interface, often as fragments of a textual whole, increase the cognitive demands on its audience. Since such a range of perspectives and media – not to mention speeds – can be accessed, exploring digital poems involves encountering incomprehensible materials that in their mode of presentation mean to have an effect on the viewer. A mix of direct and indirect communication can provide a range of possible interpretations worthy of reception, and requires new approaches to the task.

Beyond the readability of text, another important context to consider is how verbal impressions transpire on the Internet. Even someone who values visceral stimulation through language may not be prepared for the types of perusal invited by the WWW. How does reading occur on the WWW? 'By navigating or surfing,' Christian Vandendorpe has observed, 'reading is broken up, rapid, instrumental and oriented towards action.'[6] The act of reading online, in parallel to the types of writing under discussion, is fragmentary. Contents of a digital poem require more than the interpretation of mere information – putting it at odds with normal approaches to consumption on the WWW. Johanna Drucker more explicitly examines connections between print and electronic reading practices, discussing how electronic environments for reading and authoring require studying 'the basic units for viewing and organizing text/image materials, how are relations among them ordered for reading and sequencing, how are they viewed and annotated' in order to comprehend computer presentations.[7] Drucker sees graphical elements (not new to poetry) as 'not arbitrary or decorative' but as forces that 'serve as functional cognitive guides',[8] emphasizing how writers of all types seek to 'create a stream of relations'.[9] Poetry's problem, for Drucker, is the tendency for critics and readers to see books as artifacts instead of seeing them as demonstrations of 'living, dynamic nature of works as produced by interpretive acts' such as those we experience on the WWW.[10] In this processual, active context, digital poetry qualifies as less of an artifact than an active, energetic (if ephemeral) form of expression produced as a result of acts by author and viewer. 'Driving a computer as a reader/speaker/chatter/correspondent/etc. is an active thing,' writes Jim Andrews, 'and one is presented with all sorts of choices along the way be they via the hyperlink or other interaction, not the least of which is the interaction that happens just in the way we read and think and choose even when we're reading a plain old book.'[11] Establishing a correspondence between on- and offline reading practices is possible, but inevitably the demands of network-based reading are greater because its methods require actions, such as the elucidation of visual and other stimuli, adding intricacy to the activity. Beyond reading, viewers of digital poetry must absorb, interact with and interpret images, sounds and other elements; the experience demands it. Importance of text (written language) in digital poetry cannot be over-emphasized but nearly every form of electronic literature involves more from the audience than interpreting language. In conjunction with the ways they contrast general modes of consumption privileged on the WWW, the primary, integral attributes of digital poetry complicate processes of perceiving what authors intend to communicate.

Once the domain of the mechanism is understood, how deeply must a reader be prepared to analyze a digital poem? Is what occurs on the screen's stage enough to build a reasonable understanding of a work? A willingness to enter an engagement begins the process of initiating understanding; being ready to evaluate digital poems fully enhances the experience. In the following chapters – after deliberating about various conceptual dimensions within the tasks of reading – I prioritize the examination of textual particulars in completed works, illustrating the qualities a reader should expect to look for at any moment while discussing the ramifications of finite mediated properties surrounding the poetry.

What to look for

David Shepard's observation that 'a digital work is represented simultaneously by multiple signification systems and interpreted by different agents' cannot be refuted.[12] Shepard outlines three funda-mental areas of programmatic involvement in any work, which he identifies as the 'executed layer' (i.e. what is seen in the browser window, or 'the work's process as it is experienced'), the source code, and 'the execution of the code, based on the language structure'.[13] This book focuses on a digital poem's executed layer in order to facilitate instruction on how to embrace digital poetry perceptively. To understand the executed layer properly, Shepard writes, 'we must define the work's process, that component of the work that produces what the user experiences insofar as the user is aware of this process.'[14] The need to understand process in digital poetry, as demonstrated in the case studies below, necessitates discussion on expressive, technical and poetic registers. Underlying processes do inform the surface of the work through sublayers, but in fact the perceived implications of textual combinations existing on the surface adequately serve to provoke consternation and are thus prioritized in my explorations. In any event, gaining the best understanding of a digital poem requires efforts beyond what readers of print-based works are accustomed to, and alters critical parameters due to expansions of form brought on by the medium's components.

Every study involving digital literature remarks on common ground the practice shares with other genres of electronic art that combine language and code and result in the emergence of multi-modal conditions. We can, however, discriminate between genres according to form of inscription of language in a given work, as well as by any stated authorial intent. A screen, being used as a multi-purpose surface or active stage, nonetheless widens the boundaries of the literary. Computer screens 'can recreate filmic illusions, in which case the screen reflects and reinforces conventional visual assumptions', writes Hayles, adding, 'on the other hand, it can also perform as a dark window through which we can intuit the algorithms generating the display.'[15] As the surface context through which the viewer experiences content, what happens on the screen is primary. While relevant in a technical analysis, Hayles' latter point is not as important for audiences uninterested in looking/reading on this level (at least at first).[16]

On the authorial side, manipulating code, being capable of making finite adjustments to the sublayer of a work, may effectively increase its poetic values.[17] In this study, I introduce and discuss details and anecdotes regarding coding when relevant to the aesthetics of a particular work. Reading mediated works can be, Hayles writes, 'a complex performance in which agency is distributed between the user, the interface, and the active cognitions of the networked and programmable machine', requiring detailed analysis of each part of the equation.[18] Her point is legitimate, but in the interest of explaining the mechanics of digital poetry to a non-scientific audience (i.e. describing processes of reading a digital poem), I focus on a text's surfaces. Text on screen does not divulge everything, yet provides ample grist for deliberation.

Intermediation

A computer, via program/design, unquestionably has the power to produce 'a performance partaking both of the programmer's intentions and the computer's underlying architecture as symbolic processor', in which 'authorial design, the actions of an intelligent machine, and the user's receptivity are joined in a recursive cycle that enacts in microcosm our contemporary situation'.[19] To orient a theoretical framework describing digital practices that intervenes between human and machine cognition, Hayles coins the term *intermediation*.[20] The term calls to mind Jay David Bolter and Richard Grusin's concept of *remediation* (refashioning old techniques using new media), and even more directly (homonymically) Dick Higgins' experiments in *Intermedia* (crossover between artistic forms).[21] *Electronic Literature*, on the whole, profusely considers remediation in the practice of writing (an entire chapter is devoted to 'Print Novels and the Mark of the Digital'), and uses intermediation to describe a textual process (or relationship) which unites body and machine while prioritizing neither entity.[22] The experience of the encounter with digital texts, writes Hayles, 'can be understood in terms of intermediating dynamics linking human understanding with computer (sub)cognition through the cascading processes of interpretation that give meaning to information.'[23] For Hayles, intermediation introduces computation at an elemental level, although a connection to Higgins' original context is somehow ignored in the co-optation. I note it here (and return to it later) because of its relevance to understanding practices of digital literature and poetry, which seems at least as important as emphasizing issues of embodiment.[24]

Theorizing the body and technology, Hayles emphasizes the 'recursive' feedback loops integral to many works of digital literature. Her theme is relevant because many digital poems depend on user participation, which has the effect of inscribing such 'loops' in the work. Representing the feedback cycle (i.e. between self and other, body and machine) in the intermediation model foregrounds how within digital textuality's dynamics users decisively determine displayed effects. The implications of intermediation reveal some of the basic tenets of electronic literature, which do contribute to difficulties in the process and experience of reading. Hayles outlines several properties of electronic text that may cause difficulty for any reader: 'in-mixing of human and machine cognition'; 'reimagining literary work as an instrument to be played'; 'deconstruction of relation between sound and mark'; 'rupture of narrative and reimag-ining/representation of consciousness'; and 'deconstruction of temporality'.[25] In the mixture of human and machine cognition, not only do new syntactic and linguistic forms arise, but this condition may also involve issues of trust (i.e. suspicion of machine's partici-pation), or even worse, disappointment or confusion by what is contributed by the machine. Engaging with discrepant forms of information, along with the Postmodern condition requiring an audience's integral involvement in the composition of meaning to be heightened, positions the reader/viewer/user so that she or he is required to respond to an amorphous set of textual structures or circumstances. Trained audiences will be prepared to do so, yet this may be a tall order, if not a complete barrier, for those unaccus-tomed to such significant degrees of participation in a literary document.

I accept – if not take for granted – the cyborgian body-machine coupling as a given condition or apparatus, and prefer to focus on exploring the details and experience of texts proper. Observing a digital poem (on a computer monitor or elsewhere), and partici-pating in its appearance, we commune with the screen. By using the mouse or keyboard we physically interact with the machine, but I would not wholly qualify the experience as unification, and opt to qualify the encounter as watching and sometimes manipulating what occurs on the poetic stage. 'User' is a term commonly employed to describe the reader or viewer of digital work. I am inclined to describe the role variably – if a work contains mostly words I might refer to the audience as readers; if it is largely visual, perhaps viewer; if a participatory program is being used, user could be appropriate; there is no singular role we play, given the panoply of digital poetry. When an author uses software or program code, tinkering with the digital device, she or he creatively merges with electronic media in an expansive formation of expression that uses the screen as a virtual staging area. Thus my interests in the elements and results of composition supersede a focus on theoretical implications of the tools of reception.

A broadened sense of intermediation that draws in Higgins's aforementioned concept helps to understand categorically the foundations of the digital poem. Higgins coined the term *Intermedia* to identify art that combines 'aspects of two heretofore separate types of art'.[26] In his 'Intermedia Chart' (1995), Higgins graphically outlines various ways approaches to expression are combined within the sphere of Intermedia (Diag. 2.1):

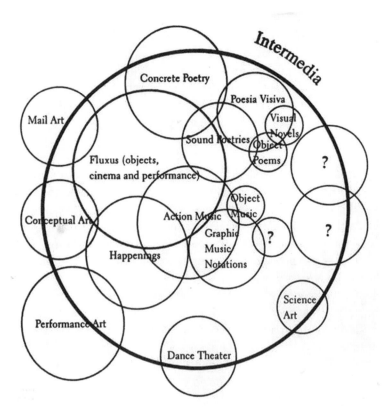

Diag. 2.1. Dick Higgins. "Intermedia Chart" (1995).

Higgins' chart provides a useful map of global polyartistry, showing how crossover between forms indicates a new model of textuality rather than the inflation of a particular form or movement. Here, an imagined plane of the chart, drawn between 'Concrete Poetry' and 'Science Art,' which passes through 'Sound Poetries', 'Poesia Visiva' and areas unknown ('?'), directly illustrates the areas of digital poetry demonstrated at E-Poetry 2003,[27] as well as in subsequent E-Poetry events.[28] Digital poets utilize pluralistic forms foregrounding visuality by letter, picture and animation. Among other effects, imagery or graphical effects populate works, often more dramatically presented on a scale larger than (or apart from) language. Beyond animation, manipulation of positive space, shading and other forms of visual texture, pictures, colours, and other factors shape digital poems. Concretism and Fluxus gave permission for poets to explore the composition of these alternative modes of articulation, and subsequently a gradual textual evolution that runs parallel to concurrent technological developments may be recognized. Intermedia as a practice became more common in the twentieth century as artists sought unexplored approaches to expression. New technologies, which evolved to include computers, also gave artists and writers devices suited to handle multiple aesthetic associations with much less difficulty. Given the historical evidence, digital poetry and other forms of electronic literature represent more than a simple technological experience – they are manifestations of Intermedia, in which visual, verbal and sonic forms merge within projected or interactive artistic structures.

Our literature and literacy have changed: previously we learned how to read in order to understand the contents of a book; with expanded modalities we need to re-learn how to experience the literary-artistic encounter.

Making Choices

A non-linear, often interactive condition of textuality presents the reader/participant with choices, through which she or he takes part in a possibly unique reading experience. As demonstrated in studies of hypertext such as *The Electronic Labyrinth*, primary challenges presented in the unfamiliar textual constructions include disorientation (getting lost in the expressive structure), cognitive overload (having to process mentally an abundance of information), and identifying 'context clues' (finding subtle, potentially valuable information).[29] The ways in which digital texts perform amplify the need for its audience to unravel artistic information in new ways.

Projected and participatory texts are performative events emerging 'across codes and circuitry within the computer and in response to interactions from the reader'.[30] One of the most important characteristics of electronic literature is that for readers 'the goal is not reaching the end ... but rather the journey itself'.[31] To surmount the variable contexts and conditions or predicaments of digital poetry, the audience needs to attend to the structure of the surface of the poem closely, be willing to inhabit a mentally active state in order to orchestrate its elements, absorb content, and build a sense of the work through the accumulation of a series of textual encounters. On the other side of the production, contemporary modes of composition offer authors an abundance of choices. Deliberate decisions faced by authors, made available amidst a wide range of possibilities, function to establish overall parameters and the ultimate appearance of the digital poem. In an interview with Marina Corrêa, Augusto de Campos observes that digital poetry opens up many possibilities for the reception of poetic forms. While acknowledging interaction may be limited to 'incisive forms, suggested by the author', he makes the point that, with digital means, poetic form(s) can be reconstructed in 'unlimited ways' and 'offers resorts of all types, even of entirely automatized compositions that are free from the personal interference of the author and the receptor'.[32] Further, celebrates de Campos, 'anyone can do whatever he wants with any text'; in the quantity of possibilities, each of which contains its own particular challenges, the genre thus 'ends up inducing a search for quality'.[33] Negotiating the combinations of creative attributes of language (and other expressive modes) and media/technology is still a relatively new task, which remains a largely experimental activity even if a classifiable genre exists. What digital poems materially consist of predicate whatever meaning(s) emerge.

In her E-poetry 2003 lecture 'It's Not That, It's Not That, It's Not That: Reading Digital Poetry', Lori Emerson openly celebrates the 'openness of interpretation and creation' in digital poetry.[34] That digital poetry conglomerates multiple artistic genres, while complicating communication, permits freedoms for author and critic alike. Yet if efforts are not made to instruct new audiences how to approach the consumption of digital literature, and why works are meaningful, a wider audience may never materialize. I aim to do so by way of intuitive, circumstantial readings of various works in the following chapters.

Possible Approaches

I once drafted, and could present, point-by-point instructions on how-to-read a digital poem. This abandoned document offers basic technical advice, such as 'reload the browser to see if the interface changes', and portrays reading as a regulated, technical act, representing only a part of the overall experience of approaching a digital poem. Classifying, in the broadest of senses, the types of receptive engagement prompted by new media forms presents little difficulty. We capture projected works with our senses and mentally process them; participatory works are mentally captured, and we respond according to sense or impulse. Critically, this means there are many possible subjective approaches and frameworks upon which to build impressions. Sensory registers at play only partly differ from those demanded by poetry. Stimulation by envisioned sight and sound effects instigate thought and mental provocations, inciting sensitivity, personal sensibility and response. The media involved create an atmosphere more expansive than that delivered by printed matter, involving eyes and ears in a different way, demanding different types of reception, recognition and interpretation. Expressive variability for authors leads to multifarious variability in terms of the way readers can approach and experience the work.

Most of the writing we encounter in our everyday lives essentially holds straightforward information. Thus the concept of literature customarily presents itself as an expressive form containing larger meanings, as something that we expect to take time to become absorbed in, perhaps in search of discovering something we are unable to see ourselves. Referencing Jerome McGann, Hayles makes the point, however, that 'meaning is not something literary texts produce but that for which they search.'[35] Thus, in preparing someone to read a digital poem, or any form of electronic writing, it is appropriate to issue a warning that 'meaning' can be pursued – or construed – in the assemblage of expressive elements, but it may not be found. For instance, observing Stuart Moulthrop's *Hegirascope*, one critic writes, 'you are now entering a labyrinth where you will not only be clueless as to where you are at any given point, but your own progression will be decided by the work itself', creating a circumstance where 'we hope to acquire enough knowledge to get a clear view of the work itself through our exploration of its maze.'[36] To invoke an age-old poetic conceit, viewers of digital poems are advised, as much as possible, to cultivate *negative capability*, a sensibility John Keats encouraged where one is 'capable of being in uncertainties, Mysteries, doubts, without any irritable reaching after fact & reason'.[37] Doing so helps any reader to pass through oxymoronic, non-sequitur associations and uncommon or unfamiliar occurrences so often found within digital poems, and may help in the process of discovering its vitality.

Communication between the vehicle and receiver in printed and digital materials differs. Conversing with Stefans about reading digital poetry at E-Poetry 2009 – where critics emphasized close readings of works produced and artists focused on developing new interfaces (including games, use of mobile devices) and creating instruments through which to conduct electronic literature – he opined that the textual scenario for digital poetry currently involves 'not comprehension but experience'.[38] Indicating that the process of reception involves more than information and data, Stefans' point succinctly pertains to the focus of this chapter and this book. Clearly, all sorts of expressive devices drive contemporary works. Many (if not most) digital poems on the WWW involve more than text – segments, sections, components that cannot be read as a poem can be read on the page.

The staging of digital poems on the screen is my starting point. Audiences for these works are required to watch closely, registering what the assemblage's language contributes. Another helpful notion worth practising is finding a 'meme montage' within the work – a combination of indicative, perhaps gleaming moments of the text revealing what stands out.[39] A text can fulfil its purpose by delivering or enabling a revelatory (transformative) experience in a sequence that can be of any length. My approach in the case studies below involves the application of multiple observatory lenses. I study the overarching shapes of each work – its forms, visual impressions, language and technology – addressing specific details of importance and what makes it distinctive. I consider particulars from within each example, in order to make assertions about a work's overall results and position in the genre's historical continuum.

Collective, or collectible, memorable moments of the poem, engraining senses – if not sensibilities – are locatable in the finest texts. In works that include media (sound, etc.), we cannot escape from its effects, which may partly distract us and must be considered as part of the overall experience of the poem. During and after projected pieces, we may consider the experience from whatever angles we individually identify. In interactive pieces, we have permission to be intuitive but can also consider organized, methodical approaches in order to absorb cognitively as much information as possible. Being patient in building a personalized perspective on a work and knowing to expect everything will not be delivered at once, as in a book, are crucial. Remaining open to any possibility is the best way to proceed in the process of perceiving digital poems.

3

Case Studies 1: continuity & diversity in online works

mIEKAL aND

Seedsigns for Philadelpho (2001)[1]
Mesostics for Dick Higgins (1998)[2]

MIEKAL aND has been an important avant garde publisher since the 1980s (Xexoxial Endarchy) and one of the first hypermedia poets in the United States.[3] He has engineered and participated in a range of digital poetry and related projects, some of which are catalogued at his JOGLARS\crossmedia beliefware site.[4] In this case study I will look at two commemorations he produced for colleagues in the contemporary poetics community.

In 2001, the Brazilian 'Intersign' poet Philadelpho Menezes died in an auto accident. When aND heard this, he composed a multiphasic Flash movie called *Seedsigns for Philadelpho*[5] (with Allegra Fi Wakest on vocals) for Menezes. This work begins with a mournful chant and epigraph and evolves into a sound and animated visual poem that introduces the narrative's characters, portrayed as letters made with seeds. The epigraph presents a quote by Menezes which addresses specific purposes of the 'Intersign Poetry' Menezes cultivated: to 'confront the realm of visual and sound effects and ... try to find ways to organize signs, in order to fill the technological products of poetry with the richness of ambiguity and complexity that signs contain when they are worked as ambivalent phenomena aimed at interpretation'.[6] Let's look at the way in which *Seedsigns for Philadelpho* matches this objective.

Graphics in aND's 'prehistoric' Macintosh-based works were rudimentary due to the software (and terminals) he used in the 1980s. Comparatively, a much less raw – i.e. a smoother and refined – realization of these effects is found in *Seedsigns for Philadelpho*, which at points embodies the frenetic textual characteristic of aND's early kinetic works, while alternatively offering a gentle and spirited testimonial for Menezes. Using Flash's capacity to orchestrate animated synthetic (i.e. processed) images and sounds, aND put into motion visual elements chosen so as to represent growth that could arise (or result) from Menezes' passing through the ideas he cultivated while living. *Seedsigns* was made with the eight letters of Menezes's name (i.e., *p, h, i, l, a, d, e, o*), combined with a polyphonic, haunting soundtrack featuring percussion and voice; the results are fluid and cinematic.[7] The audio-verbal component consists of a chant that begins with Menezes's first name followed by syllabic soundings of the letters; together these establish the work's aural foundation. A linear, projected narrative begins by introducing each of the seedsigns, spelling the subject's name letter-by-letter. As the animation progresses, words formed by these kinetic anagram- matic seedsigns ('pop,' 'pod,' 'pad,' 'had') are mirrored, enlarged and layered, as in Diag. 3.1.

Letters in Menezes' name continuously transform into a series of smaller words, each containing thematic significance – as in 'pop' above, which represents something that has suddenly burst,

Diag. 3.1. mIEKAL aND. *Seedsigns for Philadelpho.*

i.e. the poet's life. aND transmits plausible messages regarding the forcefulness of Menezes and his ideas in combinations found in the seedsigns' fleeting strings; 'had pop pad,' for example, indicates intersign methods as an explosive tableau on which a writer can operate. Concluding this phase of *Seedsigns for Philadelpho*, the letters *p, l, o, d,* and *e* enter from different areas at the edge of the screen, forming a line from which the word 'ODE' emerges and expands, at least partially indicating the work's function as an exalted song for a fellow visual poet. Multifaceted visual aspects of the animation, texturally interesting due to the character and colour of the seeds, do not impede reception here; rather, they bestow on the work a type of intrigue that responds to intersign poetry's call for ambiguity. aND's use of language (seen and heard, monosyllabic *and* polyphonic) transcends decorative identity and is, thus far, decipherable.

aND then imparts a stylistic shift through his use of kinetic permutation, which begins to obfuscate the poem. Anagrams have been a feature of poetry since at least the third century B.C. when Lycophron included anagrams on the names of Ptolemy and his queen, Arsinoë, in a poem about the siege of Troy entitled *Cassandra*.[8] Anagrams have served various functions in poetry, particularly to cloak names and other types of information. However, aND's use of anagrams

Diag. 3.2. mIEKAL aND. *Seedsigns for Philadelpho.*

Diag. 3.3. mIEKAL aND. *Seedsigns for Philadelpho.*

is far more complex – and certainly more demanding on viewers – because permutations occur quickly. Four letter words appear on four lines, set up on a loop, each of which repeats a different number of times, atop a shifting background pattern of visually processed seeds, enlarged and green in tone, as in Diags. 3.2 and 3.3.

Rapid transitions between words pass too quickly to be read conventionally.[9] Given the number of words in each line (6, 3, 5, 4), 360 word combinations are possible, and while every combination is not presented, an exorbitant amount of visual activity occurs during this approximately eighteen-second section. By varying the number of words in the loops of each line, aND enables more deviation of appearance in the combinations appearing on the screen. A viewer has to register words, mentally connecting those that stand out to form a poem; along the way the viewer may discover a type of lyric within them, given the possibility of rhyme (as in 'dial' and 'pile' in Diag. 3.3). Words chosen by aND are poetic in that they provide openings through which the receiver momentarily transfers perception or consciousness from one place to another. Positing the poem as homage, we can see how language begins to function in this capacity by virtue of the words chosen for the opening line, e.g. 'heal,' 'hope,' 'held,' 'heed,' 'hide,' 'head'.[10] Screen captures, as above, can be used to isolate sensible fragments of poetic speech, e.g. hope/deep/loop/pile, Diag. 3.2, which perhaps self-reflexively reveal instruction as to the work's meaning (or what messages it contains). Invoking 'hope', for example, we cannot know the author's exact connotation, but several come to mind. Is it inspiration – finding hope in the idea that Menezes' artistic and theoretical work becomes known and/or more widely realized? Is hope itself a deep looping pile? Should we look for hope in aND's poem, which gives back, in gratitude, for Menezes' efforts?

aND's composition serves to transform grief for loss into something positive. In other words, 'Heed' (Diag. 3.3) represents a call to listen to, perceive and learn from the departed subject. Words selected have an overall poetic function through their suggestive (but not conclusive) properties, method of presentation and level of variability. Viewers familiar with Menezes' work may contemplate connections between words chosen, how they are arranged and Menezes' genre-fusing theories.[11] Rather than incorporate archetypal

forms of intertextuality, Menezes promoted a 'network of connections based on technological links made available by hypermedia;
a network of associations set up between the data of the poem,
which refer to each other'.[12] He practised this in a context that privileged suggestion over explanation. aND, by mechanically associating
words (anagrams) with diverse yet pointed connotations in combination with radiant, strategic images, clearly draws from these ideals
while orchestrating *Seedsigns for Philadelpho*.

Drawn into the poem by language and other effects, viewers
capture fragments, possibly to the point of being temporarily hypno-
tized by mesmerizing sounds and flashing, layered imagery. Although
different interpretations of the combinations of words are possible,
any viewer should begin to grasp a sense of the poem's scope after
watching it several times. A direct message in the arrangement of
letters is sometimes unclear if not insensible, but the poem establishes its domain through its mode of strict order, reflecting the
way nature and organic communication offer creative possibility.
Perhaps aND could have fashioned an aleatoric setting or a script
that randomized the permutation, thus making them different each
time, but that may have subverted his intention to pay homage to
Menezes.

Seedsigns for Philadelpho is a classic animated digital poem: a
linear presentation of media elements synthesized through Flash.
At least two of aND's authorial approaches make the work difficult
to read: the poem does not follow ordinary syntactical arrangement
and the anagrams appear in series of briskly changing lines. The
words assembled do not straightforwardly elicit meaning, and speed
of presentation inhibits the ability to read with ease. Its hand-spun
elements – combining organic iconography and digital technology –
propel and elevate the aesthetic qualities of the poem, rendering rich
poetic principles. A grouping of letters (individual units in appearance
and sound) become multifaceted characters in the work, collected
and then energetically recollected to deliver expression. The poem
sustains itself through a type of recycling in its verbal components.
Working directly with a piece of hardware (a scanner) and carefully
arranging hundreds of seeds into letters, aND regulates shape and
colour so as to present visual contrast, texture and change. Formal
elements in the work – a coded articulation and an articulation of

language – contain poetic symbolism and offer a spectacular setting for the solemn yet also frenzied verbal manifestations found within.

Now let us look at another commemoration created by aND, *Mesostics for Dick Higgins*. In contrast to *Seedsigns for Philadelpho*, aND's *Mesostics for Dick Higgins* is monophasic; its sensorial staging is homogeneous throughout. The poem's rendering is by design consistent, never varying from its jarring kinetic depiction of language. However, in spite of this ostensible one-dimensionality, aND achieves artistic valency through his choice and method of representing verbal content. Fragmentary linguistic delivery and overall technical processes combine to complicate its reception.

Dynamically, despite the differing compositional approach, computational methods and appearance, *Mesostics for Dick Higgins* corresponds with *Seedsigns for Philadelpho* in two respects. As a foundation aND implements automated loops of words, the compilation of which produces mesostic poems commemorating Higgins.[13] aND's poem, mined from Higgins' book *foew&ombwhnw* (1969)[14], arranges words into successive, incessant eleven line formations (a line for each letter of the subject's name), as in this example:

subsiDized
            Is
        reCultivating
    worK
someHow
        antI-religious
        judGes
    brinG
        utilItarian
meaninglessNess
            juStified.[15]

In this poem, words shift tectonically and almost uniformly. As in *Seedsigns for Philadelpho*, aND programs looped lines to appear in the same ordered pattern for a set amount of time when the piece is activated, in this case for one second, according to the code.[16] However, the timing sequence becomes skewed or disjointed because each of the eleven lines constantly opens and closes, and the volume of processing necessary at every moment varies the speed of presentation depending on the local computer's capabilities. Delays of computational processing create an ongoing visual stagger, practically ensuring that output on the screen is never quite identical; it is always close to being co-ordinated but is not. Without concretely synchronized lines, text appearing on the screen – which passes too quickly to read any single iteration as a whole – steadily varies. A reader faces a constantly shifting poem that eludes full capture, unless s/he preserves a screenshot of it.[17] Since aND does not present the poem as a series of still images, however, we know the work is intentionally bombastic, purposefully inundating the viewer with a panoply of Higgins' words.

When a fragment of the poem is captured, readable passages emerge – abstract on the surface and in need of creative interpretation. Engaged readers can take time and piece together, word-by-word, overall results of the pre-structured but automated utterance. Even then, however, specific meaning may be elusive. Analyzing the example presented above, for instance, separate sensibilities can be cobbled together: that assistance (subsidization) of some sort serves to further promote (recultivate) labour, and that people without something to believe in (anti-religious judges) bring useful meaninglessness. Establishing connection between the two parts involves effort. The poem turns at 'someHow'; fusing the seeming disconnection between beginning and end is the reader's responsibility. How does the partly happenstance conglomeration of Higgins' vocabulary form to say something about people who are not religious? Do 'antI-religious/judGes' justify 'utilItarian/meaninglessNess'? Is 'utilItarian/meaninglessNess' something to be celebrated, or not? Significant work is already required of readers who get to the point where they ask such questions and they must exercise further poetic interpretation to proceed. The fact that this poem can spark such interpretations with words removed from their original context demonstrates the provocative qualities of the original choices and source of language, fortified by the digital poet's treatment of them.

Many different, fleeting sets of verbal juxtapositions occur. Some combinations are less comprehensible, and reading the poem as it runs prohibits such clinical availability (fixity) of any emergent text. Regardless, a second example of a preserved passage from *Mesostics for Dick Higgins*, with an entirely different vocabulary, also succeeds in portraying an identifiable message.

```
galvanizeD
          fIlter
          Cools
          Keyboards
          tHis
          Itself
    stockinGs
      strinGs
          lIttle
        thiNgs
    bodieS.18
```

In this instance, a strengthened ('galvanized') 'fIlter' (an entity open to interpretation but that applies to digital treatments of matter, or something that removes unwanted qualities) alters the heat of activity ('Cools/Keyboards'). This makes it part of a larger whole (including 'lIttle/thiNgs'), thereby situating humans ('bodieS') in a larger context. A reader could, among other possibilities, interpret the lyric as a self-reflexive address to the act of programmed composition; in other words the presentation involves concept, décor, instruction, as well as a mental and physical presence.

The likelihood of each reader taking something different away from this hyperactive poetry is great. Variant themes can emerge from the unfixed reconstitution of words from Higgins's book, as in the samples above. Permutation enables different voices, topics and styles of expression to emerge, thereby distributing the original content into new contexts. As I asserted in *Prehistoric Digital Poetry*, 'selecting appropriate input text is the most important element in the process of pronouncing meaningful expression'.19 Therefore aND, as selector of the words from the original source, accomplishes something impressive here. Not only has he orchestrated a program that successfully imparts an unusual structure to a kinetic digital poem, but he does so in a manner that produces multifaceted output despite containing a relatively small database.

Certain features that aND opts to use contribute to making *Mesostics for Dick Higgins* an interesting and unusual textual event. aND has carefully prepared an HTML page for every word in the poem (187 words in all); HTML inserts words into a meta-framework (vertically oriented slots, one for each line),[20] appropriately placing each into the overall structure, i.e. the HTML framework containing 11 lines. Despite rigid and ornate programming, aND manages to promote a sense of looseness and ephemerality in what appears on the screen. Its kinetic qualities prevent reading the poem in any conventional sense (no more than three or four words can be read at once); in essence, chaos prevails. Words appear and disappear at a rate guaranteeing some readers will experience apprehension.

However, by suspending ordinary reading practices and remaining open to what the poem presents, one discovers several ways to absorb the poem's literary and artistic traits. Any of the individual lines, which take on Möbius properties, can be a focal point and entered *in medias res*. For example, concentrating on the first line only, a reader can absorb the sample, 'Doing/passeD/Degrees/closeD/heaD,' without difficulty. Mesostic form, by nature, minimizes alliteration and enables a varied line, which in this instance might be read as a complaint about an academic experience. At any moment, a reader can shift focus of concentration to a different line and continue reading as such. This approach may be horizontal at first (reading across a line), and then vertical (proceeding up or down to another line). Reading one word (one line) at a time and immediately moving vertically through the poem is another way to process the language of the work. Then, at the last line, the reader can return to the top, start again, and will encounter a different eleven-line 'stanza' when repeating the process. Finally, a reader can also develop a completely random or haphazard reading style, picking out any word from anywhere on the screen and mentally cobbling together a poem along the way. Given the fact that less than two hundred words appear in all, the possibilities are not endless; the words become familiar after spending several minutes with the piece. Nonetheless, aND presents great variety of expression, within which narrative progressively shifts.

Mesostics for Dick Higgins effectively distorts fragments of one artist's work, appropriated by another, who – in the experimental tradition – reassembles them in a different but relevant context that serves to bestow attention to the inventions of both. aND's production is less subtle than Higgins' because the media he has access to enables a more spectacular noise. Yet its *gestalt* remains true to Higgins' lack of subservience to singular forms.

With *Seedsigns for Philadelpho* and *Mesostics for Dick Higgins,* aND delivers unconventional projected works as a linear experience, in the tradition of humanist commemorative poetry, which certainly involves processes not new to poetry; both could have been produced, in somewhat different forms, before the computer age.[21] His presentations exploit videographic possibilities for literature, thereby illustrating how contemporary methods are more fluid and capably deliver content with immediacy. Both works eschew ordinary syntactical arrangement and their speed of presentation prohibits a typical reading experience. Processes aND uses in composing language are not new, but the way he projects output has a heretofore unseen visceral effect. By using hypermedia, aND gains the ability to shuffle language hypnotically, potentially even putting viewers into a trance state, through which something unpredictable might occur, such as transformed consciousness. Quick renderings of words contained in the output do not intend to appease readers seeking conventional writings, who can download aND's books from the WWW.[22] Rather, poetic force rolls and varies, providing numerous ways for its viewers to absorb what otherwise could be static material.

Jim Andrews

Stir Fry Texts (1999-2000)[23]
Arteroids (vers. 3.11, 2001–7)[24]

Jim Andrews, like others whose works span the divide between past and present modalities, has produced a series of dissimilar digital poems on the WWW.[25] Given that he has administrated a website called Vispo (a nickname given to visual poetry) since 1996, it is unsurprising that visual characteristics unify his disparate works, all of which require viewers to mediate the transformation of given materials.

First we will look at *Stir Fry Texts* (1999-2000). As the cooks among us know, 'stir fry' refers to a cooking process wherein many ingredients are mixed together with oil in a pan and cooked at high temperature on a stove. At any given time in the process – from cooking to presenting to consuming – the ingredients get mixed up so that some become hidden underneath. Likewise oil fuses the flavors, keeps the ingredients from sticking, and distributes evenly the essence of each. Stir fry dishes include many flavours and ingredients, making it difficult to taste each one of them simultaneously. When stirring the contents, momentary glimpses of what is beneath become possible; an overall sense of the dish's structure is gleaned. This metaphor aptly describes the style of Andrews' *Stir Fry Texts*. Beneath the engineered surfaces of the poem, each of which performs by shifting within a singular graphical context on the screen, Andrews poetically shapes and interconnects multiple layers of text in an HTML environment through Cascading Style Sheets (CSS), JavaScript and tables. He interactively transmits many different texts from within a single visual framework, fused by the programming's 'oil'.

In *Stir Fry Texts,* mouse-activated combinations (movements) summon jointed or disjointed narratives. While interacting with one section of the work, 'Divine Mind Fragment Theater', I am suddenly, surprisingly – completely irrationally – reading a portion of Joseph Weizenbaum's *Computer Power and Human Reason* about Emerson's use of Language Poetry. This is glorious folly, which at the very least provokes one to consider what Emerson would have thought about Language poetry.

In this project Andrews synthesizes techniques found previously in text-generators with techniques used to create 'virtual objects' within the realm of hypermedia forms. *Stir Fry Texts* consists of seven distinct poems.[26] Each is participatory: only by engaging with the work does its content become apparent. By moving the mouse over text on the screen, viewers activate alternating layers of text virtually bound together. Each distinct layer of text is colour-coded, so output presented on the screen at any given time features passages in multiple colours, which may be perceived as different 'voices' within the poem.[27]

At the outset of the series' first piece, 'Log', which features texts provided by Brian Lennon, five passages appear as white text on black background, divided into sequential time intervals in red.[28] Thus it seems the viewer encounters a linear account of reflective moments occurring between '04:51:00' and '04:55:00' (SFT) – not a reflection of *real* time, but rather Stir Fry Time. When the viewer begins to navigate through this material, the screen content instantaneously transforms. Two additional layers of language appear in different shades of grey and, when activated, mix with the original text. The result is that the poem's surface, including the initial timeframe, is altered by the additional layers. One of these layers is written in the first person, suggesting a narrative. Dramatic opening lines in 'Log' illustrate these characteristics while introducing technology as a theme in the poem:

> 04:51:00 –
> I split the full scale mark – I choked and drooled – I was
> memory bound – I muted analysis – I thought I'd start –.
> 04:52:00 –
> I'd been aware – I'm dead I'm dead – I'm standing up –
> I'm what you want – I programmed it –.[29]

In this excerpt, Andrews simultaneously presents a sense of containment ('I was/memory bound') and a sense of satisfaction ('I'm what you want – I programmed it'); the narrative addresses a type of struggle for life. As layers of text continue to intermingle, a sense of action and drama is perpetuated. Here, digital text reflects, or more figuratively represents, human anxiety. Expression, at designated times, is collapsed or reordered, sometimes disappearing altogether. Because of the declarative, first person statements, the output retains a sense of personal urgency and anxiousness. This approach continues in the excerpt below, which draws from each of the three base texts (distinguished by differing formats below rather than by colour as in the original):

> *vox coriolis* –
> I split the full scale mark – <u>made contact –</u> I was memory
> bound – I muted analysis – <u>milled out –</u>
> *it was the light* –

I'd been aware – I'm dead I'm dead – I'm standing up – it
rose – I programmed it –
electric vex –.[30]

Studying the two examples above reveals the work's main purposes. Dissecting the second iteration of the poem, readers will note some retention of language and that the line breaks and poem's shape have shifted. Visual flexibility abounds, signifying semantic shifts. A telling characteristic of this work (and perhaps of digital poetry) becomes evident: its self-reflexivity. The second passage offers a possible clue as to the apparent crisis in the poem: it may be, in fact, the mechanics of the poem – or this type of poem – itself. Certainly, the dynamics of the poem reflect a condition wherein the voice ('vox') of the poem personifies the *Coriolis Effect*, which in physics describes the deflection of objects when viewed from a rotating reference frame. Since *Stir Fry Texts* gives its readers at least two points of reference, the poem's narrative ably reflects its technological apparatus. Consequently, panic may be replaced or accentuated by successful action, e.g. 'made contact', 'milled out', and transcendence or understanding, e.g., 'it was the light', 'rose'; – in other words, its drama becomes more of a classifiable puzzle ('electronic vex') than a calamity.

Spliced juxtapositions occur throughout *Stir Fry Texts*. In 'Log', a sense of frustration with technology is sometimes reflected. Take for instance the lines that appear upon further interaction, which indicate a problem with code: '*nothing began* – was late – /*parsed element 01'*. Since many different instances of the same textual space are possible (e.g. the same space of the screen can be quickly altered to read 'changed my life – I was amazed/ *parsed element 01'*), radical shifts transpire and problems find resolution; it soon becomes clear that finding a singular, specific polemical messages from the work is not feasible. Internal, mechanical changes in combinations of language transform any meaning that arises.

Raymond Queneau's *100,000,000,000 poems* should be con-- sidered a predecessor to *Stir Fry Texts*. Queneau's book contains sliced pages that enable readers to transpose lines of a series of sonnets to create new combinatory poems. The series of slots that morph within Andrews' poems resemble the layering of different pages that readers conduct in Queneau's sonnets (also available in interactive forms).[31] Combinatory works represent advancement for Andrews, who in the pre-WWW era made static work. *Stir Fry Texts* – which Andrews associates with previous experiments done using 'cut up' techniques – makes use of technology and media in a way Queneau could not: the poem and words that comprise its lines visually expand and contract given the viewer's movements of the cursor, and layers of text alternate quickly on the screen. These features, writes Andrews, present a 'spastic interactivity' for the viewer to work with.[32]

Other sections of *Stir Fry Texts* employ this technique with different outward effects and results. 'Spastext' (1999) uses five layers of text, written in paragraphs, as a platform for a manifesto about artistic self-empowerment.[33] Unlike 'Log', which is peripatetic but readable, 'Spastext' utterly garbles syntax when fragments of multiple texts appear at the same time. In most of the *Stir Fry Texts* rules of grammar are, following experimental custom, ignored. As individual layers break apart and then merge together with other layers, however, powerful statements form, which is a consequence of language previously exposed by Dada artists. 'Blue Hyacinth', created with Pauline Masurel, unifies four separate vignettes using the 'Stir Fry' process. Each strain tells a different story, which come together to form interwoven non-linear passages that do not reify the given individual texts. The original prosaic stories become indiscernible in the disjunctive heaps of text fashioned by the viewer/ participant, and several different meanings for the phrase 'blue hyacinth' are offered, e.g. one is the name of a horse, one describes a woman's voice. These authorial choices aim to create multiple narrative directions. Predominant use of the pronouns 'she' and 'I' enables diverging strains in the narrative to meld into a sensibility, although, as throughout *Stir Fry Texts*, viewers must always suspend the need for accurate punctuation and proper adherence to the standard rules of writing. The 'Correspondence' and 'Divine Mind Fragment Theater' sections particularly embody the surrealist practice of 'exquisite corpse' writing; Andrews blends selected writings to achieve hybrid articulations.[34]

Stir Fry Texts are intertextual (if self-contained) and multitextual; they are simultaneously singular and multiple. Users who familiarize themselves with the individual layers and tinker with what appears on the screen essentially co-create customized – if ephemeral – output, though ultimately these pieces resist offering users total control. Dreamlike, histrionic and sometimes even funny scenarios emerge; these are not to be collected and documented but destined to remain electronically open and extend into new dimensions. Viewers follow the continuum, directing it systematically or accidentally. Results of these experiments resist smoothness in terms of readability, but effectively produce text embodying multiple voices and viewpoints.

In this work, Andrews explores issues of authorial control and novel writing processes. Beyond its awkward yet engaging qualities, the 'stir fry' process makes a noteworthy point: writers cannot avoid re-writing the words of others. As Memmott asked in his contribution to 'Correspondence': 'Is a text ever NOT a cut-up of sorts?'.[35] Is a singular voice possible to identify in any document? What kinds of statements are made without using the words of others? *Stir Fry Texts*, while non-graphical, are textual spectacles illustrating the depth of digital writing surfaces.[36] Becoming accustomed to the interface's spasticity, dedicating time to understanding emergence of the sub-layers of text, viewers can create multi-linear texts through interactive participation with Andrews' mechanism. Andrews has carefully selected texts that 'collide and interpenetrate in an interesting and hopefully startling and even enlightening way'; he has chosen unique methods of splicing disparate materials together.[37] Using instantaneous hyperlinking and colour-coded information to delineate various layers of text – in themselves not new techniques employed by digital authors – Andrews' original combination of technological application illuminates an historical process (i.e. cut-up technique) while at the same time making thoughtful pronouncements through unconventional narratives. A special type of interactivity devised by Andrews creates the character of the piece(s); the unorthodox response of text to mouse movement allows a certain degree of viewer 'control' but also imposes a definite degree of jurisdiction by the machine.[38] Users intervene, but constraints ordered by the program are palpable. Andrews creates through the work's programmatic 'oil' an invisible character that consistently exerts its own behaviour, evenly distributing or scrambling the language in its database.

'Stir Fry Texts', following Queneau, essentially creates collages of passages – fragments – of text written by individuals and arranges them so as to emit, with balance, a potential story or message. Yet the process of manipulating the hierarchy (order) of lines is instant, kinetic and difficult to control precisely – programmatic features that make a profound impact on the delivery of the literary product. The processes may not be new, but the product is dissimilar to any print-based work. As in Queneau, many iterations of output arise, but unlike the experience of absorbing a book, viewers interacting with *Stir Fry Texts* efficiently receive a more refined sense of the possibilities held by permutation. A reading of multiple texts occurs fluidly, in a different order, for everyone. We do not need to read all the possibilities to understand the results of its qualities. The poem emerges, as always, through the processing of text. Beyond discovering the types of texts the piece generates, readers develop a concern for digital textuality and how authors involved with the *Stir Fry Texts* approach these issues.

Let's return our attention to another work of Andrews', *Arteroids*. This digital work, as described at the outset of this book, creatively re-versions the once popular Atari video game *Asteroids* (1979).[39] Describing it as a 'shoot-em-up poetry game' and 'battle of poetry against itself and the forces of dullness', Andrews does not create an authentic modification to the original code but rather adds his own (adjustable) poetic content and features to the general design of the game.[40] Available in English and Portuguese, *Arteroids* features two different modes, 'Game' and 'Play', which resemble each other in appearance but differ in functionality.[41] In both, words (sometimes phrases) and alphanumeric symbols – the arteroids – emerge from the edge of the interface and drift across the screen. Some arteroids gravitate towards the user-controlled object/target, dubbed the 'id-entity' text by Andrews, signifying the 'being' within the game that embodies a personality structure containing the basic drive to play, or negotiate and survive, the verbal onslaught. Arrow keys move the id-entity text, and pressing the x key on the keyboard enables the user to 'shoot' virtual bullets at the arteroids. Players determine progress within a level by reading the 'meanometer' (i.e. the meter of meaning) appearing on the right of the screen. Verbal density of the poem correlates directly with the level indicated by the meanometer; the goal is to shoot enough texts to move on to the next level. Scoring (the actual numbers) is not arbitrary, but does not appear as play occurs. An unstated formula for scoring, 'based on accuracy and time', has been devised by Andrews, but scores are not revealed until after play has ended.[42]

In the following examples (Diags. 3.4 and 3.5), taken at different phases of a game of *Arteroids* in Game mode, position of the id-entity (i.e. 'poetry', in red) is roughly the same, but the background, configuration and density of verbal information surrounding it completely differ.

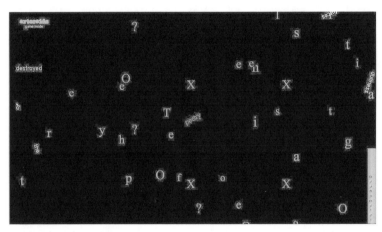

Diag. 3.4. Jim Andrews. *Arteroids.*

Reading the above diagram while the game transpires is possible, although distractions occur due to frenetic motion and abundance of activity happening on the screen. Reflecting on this moment of the poem via screenshot, scattered words and letters reveal several insights about the poem and Andrews' project. The arteroids 'destroyed' (at left) and 'dullness' (angled at right), thematically implanted by Andrews, remain intact. A few small words (e.g. 'of', 'is'), larger Xs, Os, ?s and other letters representing remnants of words appear prominently but do not overtly connect in space – they add content but not necessarily as formed language. Coincidentally, 'e poetry' appears at the centre of the screen, serving to locate the event within a context of E-Poetry (a name for electronic poetry), which would not have occurred had the id-entity been moved elsewhere. As in *Stir Fry Texts*, small actions, reactions or user movements play large roles by altering the content and arrangement of words on the screen. Amidst the fragments, or fallout resulting from the player's actions, the phrase 'destroyed is dullness of poetry' (or alternatively 'destroyed is dullness of e poetry') directly announces Andrews' programmatic intentions, as indicated by his stated aspirations for *Arteroids*.

In a later instance of the same game (Diag. 3.5), the phrases 'created poetry' and 'destroyed poetry' emerge, as does the vertical construct 'all/poetry/destroyed/poetry'.[43] These and other snippets of text (among other visual debris) may be registered or discerned by the user while playing the game.

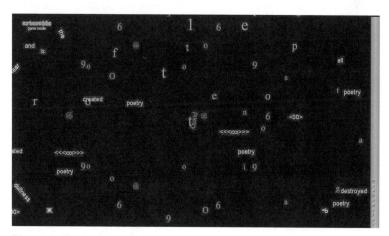

Diag. 3.5. Jim Andrews. *Arteroids*.

Ideally, in Game mode, the user both creates and destroys a severely constrained poem, constructed amidst a range of verbal possibilities, concurrent sounds[44] and onscreen kinetic activity. This surplus of stimuli is part of the poetry, designed to enliven it as a *played* multimedia form. Words (arteroids) disfigured by the user are put to alternative uses. Their function within the poem is theoretical and decorative; small happenstance words (e.g. 'toe', 'to' and 'tot' in Diag. 3.5) might be assembled and used to construct minimalist, compact lines of poetry containing subjective or objective meaning. The perceived line 'to toe poetry', for instance, could be connected

to the purposes of *Arteroids*. Andrews makes poetry while experimenting with, and inventing, new ways to make it. For some players
this type of broadcast and reception of text may be important; others
will focus on staying 'alive' in the game, reaching the next level, and
earning a high score. With the id-entity under siege, reading the
quickly moving arteroids, particularly in Game mode, is not always
the priority (and is easier during early stages/levels of an encounter).
Even in the best of circumstances, the need to respond to multiple
textual elements severely curtails the player's ability to read text,
making possible only brief encounters with poetic language. This
characteristic is among the many in *Arteroids* that position the work
in contrast to types of poetry containing lengthy, heroic, ornate lines.

In Game mode the user-driven id-entity of the poem (i.e. 'poetry')
reconfigures the text of the arteroids and/or is eventually demol-
ished. As arteroids appear, the id-entity transforms them while
being manipulated to avoid them; as players reach different levels
of the game, the speed of movement and level of difficulty of these
tasks increase, making any type of reading a severe challenge.
When explosions of arteroids occur, letters (converted into other
letters or symbols) spread out: sometimes in shapely patterns, and
at others dispersed at random. Detached and repositioned letters
do not move on the screen and gradually disappear from sight. This
shifting background provides part of the poetic (textual) experience,
supplemented visually by intact floating arteroids, and – perhaps
centrally – by the immutable but moveable id-entity. The sum of this
predicament of textuality, including sound and motion, in a nutshell
stalwartly reflects the conditions of digital poetry. Full of variation and
mediated possibility, words and symbols perpetually – sometimes
interactively, sometimes randomly – reconfigure and replace one
another.

In Play mode, *Arteroids* offers users an opportunity to alter and
enlarge the scope of the poem. The arteroids and id-entity may
be adjusted through a function called 'Word for Weirdos'.[45] Colour
and size cannot be changed, but users have the option of choosing
default texts (by Andrews, Helen Thorington or Christina McPhee),
editing the given texts, or replacing them with their own words,
i.e. composing new texts to be used as input. When one glances at
the diagrams of Play mode seen below (Diags. 3.6 and 3.7), one of

the first noticeable differences between them is that the id-entity has changed from 'poetry' to 'desire'. Although this id-entity can be changed to anything, Andrews's decision to use 'desire' as his default setting reveals an essential difference between Game and Play: users of Play mode want something else from their experience, something beyond competition with self and others, speed and proof of dexterity.[46] Recognizing creative possibilities for the gaming model, Andrews engineers in Play mode an iteration of *Arteroids* enabling users to communicate and save 'creative texts/messages', thereby imparting a scenario in which 'life at risk is choosable' – meaning the game will continue as long as the 'meanometer' is not empty.[47] No matter how many times the id-entity is destroyed, it re-appears on the screen and play continues. Play mode extends textual possibilities within *Arteroids*, potentially offering more versatility and nuance.

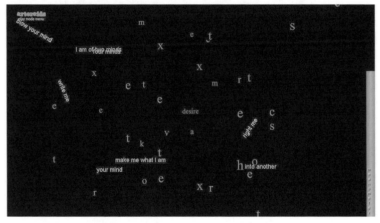

Diag. 3.6. Jim Andrews. *Arteroids*.

In effect, users now play by conducting the arteroids, by modifying both the initial texts and also what they become after they explode. Using a default setting such as Andrews' 'Writing(arteroids)', unified thematic ties between 'before' and 'after' states of text do not always appear, but users definitively establish before and after associations.[48] In one set of default configurations, the line 'two texts' explodes into the binary numbers, 0 and 1; in others, 'I am an enigma' becomes question marks, and 'make me what I am'

forms the word 'Blue', which is the colour of the text. By incorporating more texts and strategically destroying particular texts, some
control over the final output (i.e. what the transitory poem consists
of) may be attained. These effects widen the boundaries of readable
and viewable content. Since players cannot adjust sounds and
colours, *Arteroids'* overall aesthetic and functionality remain intact.
While the appearance is similar in both modes, in Play mode users
may configure parameters with much more intricacy, including the
velocity of the arteroids, the 'friction' of the id-entity (i.e. how quickly
it stops), textual density and size of explosions.[49] Each device serves
to alter the parameters of experience and what *Arteroids* emits as a
whole.

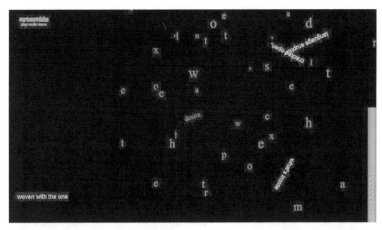

Diag. 3.7. Jim Andrews. *Arteroids.*

Diags. 3.6 and 3.7 are moments captured from an encounter with
Arteroids in which Andrews' 'Writing(Arteroids)' is the input text. Text
samples in this segment address 'language eruption', 'mind rupture',
and 'meaning haemorrhage' (in green), and the weaving of texts
('sextext') (in blue). Innuendo to physicality and writing as a sensual
act are at the core of the sublayers, though they are not overbearing
as such in this example (Diag. 3.7), which reads (following a
counterclockwise spiral): 'desire/woven with the one/what's inside/
cracked open,/language eruption'.[50] This sampling reveals how input
materials, in form and expressive communication, can be subtly
modified through creative processing and excision during play. We

see again a self-reflexivity, a projection insinuating or embodying the type of poem being made. This is, however, not a permanent condition of the work because input texts and letters appearing on the screen via the arteroids may be adjusted and can contain any words the user keys in.[51]

In digital poetry, practically any structure can be used as a model for design. For example, Rob Swigart and Jena Osman, to name just two, have used the Periodic Table of Elements to structure poems. Larsen (discussed later) uses the sculpture of Michelangelo's David, and Jim Rosenberg (discussed later) invents his own visual conception as an organizational mechanism. That Andrews chooses a video game as the model for a poem is not surprising, given the popularity of the activity in mass culture, thus making it a large target for appropriation. Most impressive, given the programmatic demands of building such a complicated device, is the will digital poets have to embrace difficult tasks, clearly demonstrated by Andrews here. These programmatic demands overtly correspond to the difficulties experienced by readers. Andrews makes literature into a game that atomizes and vaporizes texts. In the process, given elements are reconfigured in an extraordinary, if ostensibly fragmented, manner. Because of the level of integral interactivity required, absorbing all the elements at once presents challenges. In Game mode, users are constantly under attack and must shoot and destroy or ably move around the screen in order to continue playing; the amount of attention users can devote to all the details on the screen will not be comprehensive. Phrases can be read when the speed of the arteroids is slow, but not all the details within the materials – especially the fragments (letters) resulting from broken arteroids – can be fully absorbed. When analyzing screenshots taken of games in progress, one can begin to discover potentially meaningful associations between the elements, but this does not occur in the midst of the interaction.

Arguing for Andrews' approach as an essential direction for digital poetry given the degree to which he makes demands on players, would be difficult. Moreover, his project – making poetry into a game – could be contested by both traditionally-minded readers and serious gamers.[52] Yet beyond the enormous labour involved in the work (the version of *Arteroids* reviewed here was refined for more than six years), Andrew's creation of a tool that appeals to a mass audience by virtue of the idiom it embraces is admirable; this approach is likely to be developed further by artists and educators who wish to acknowledge shifts in literacy. Because verbal output is visual and sonic as well as alphabetic and resists deep coherence in a conventional way, the chances of *Arteroids* gaining a massive audience are small. Average readers and game players are not ready to embrace this degree of abstraction and will be challenged to find a way to read output fluidly. However, *Arteroids* serves, with its range of aesthetic qualities, to indicate potential modes of poetic (literary) communication, and begins charting such artistic territory.

Gaming and literary art may not be a perfect match for each other given the multitasking required, but attributes of certain games are ripe for poetic co-optation.[53] Another example of how an archetypal video game is used as a basis for a conceptual digital poem is Neil Hennessey's *Basho's Frogger* (2002).[54] Hennessey co-opts his motif from the Konami arcade game *Frogger* (1981), but instead of creating a game that the user can play and win, he designs one that cannot be played successfully. Using Java, a device is built – using a replica of the *Frogger* interface – that ensures that the 'player' is immediately killed and scores no points. When the game is over, a pre-programmed 'score ranking' screen appears, containing zero points for players 'pnd', 'frg' and 'plp'. In this work, Hennessey makes a direct association to a piece of historical literature, specifi-cally a haiku by Matsuo Basho: 'The old pond/A frog jumps in/The sound of water.'[55] Content of the haiku provides the thematic context for Hennessey's piece, which suggests a game but does not reify its playability. Works such as this and *Arteroids* clearly represent a different kind of poetry and authorial intent, as indicated in 'tips' provided by Andrews: 'Turn out the lights! Turn up the sound! Throw away your preconceptions about poetry!'[56]

Andrews builds a tool that creates on-the-fly visual poems. When the word 'poetry' breaks apart to form a circle of *o*'s on the screen or when another word scatters haphazardly, decorative, textural and other effects are inscribed into the game-poem. Users experience sonic fragments and bits of text, which could harmoniously reform into lines resembling conventional formations but do not.[57] In his introductory essay, Andrews offers direct insight about what he communicates in the project. He writes: '*Arteroids* is about cracking language open' – a point which can be taken literally – and 'gets its primary meaning from the meaning of the words to text as meaning via sound, motion and destructive intent.'[58] The id-entity, as a warrior for art, transforms words as they normally appear, spreading them out across the screen so players can potentially create their own words or mental states with them. In order to establish new forms of poetry, Andrews makes a gesture towards destroying the old – which his device capably does – as an entertainment into which meaning may or may not be read, i.e. players might not prioritize reading, in favour of playing the game. When the danger of being eliminated from competition is not an issue (Play mode) and the work can be enjoyed without unceasing involvement, *Arteroids* underscores tensions between poetry and game, art and game, and poetry and play without resolving them. Instead, Andrews dramatizes, highlights, and makes evident these relations – presenting them within a type of lettrist cinematographic experience whose appearance, i.e. content, can be amended by the user/player at any time.

The presentation of poetry in *Arteroids* is certainly unlike anything a print-based author could achieve. Particularly within its Play mode, where readers can customize all textual content and strategically participate, authorial processes are novel. Few examples of electronic literature have ventured to offer such a Futuristic, open and flexible poetic system. How often are readers required to destroy words and phrases to make a transitory poem appear? Although the output will always resemble visual poetry, and this work can be said to partake of that tradition, Andrews ultimately surrenders control of composition to his program and user.

John Cayley

Wotclock (2005)[59]

Cayley, whose initial processes and HyperCard works are highlighted in *Prehistoric Digital Poetry*, uses new methods on the WWW to produce his innovative 'machine modulated' poems, in which multiple texts programmatically merge and interact to create new output.[60] He actively uses the WWW to publish digital poems and performances, effectively extending his early Indra's Net project to the network itself.[61]

Wotclock, featuring panoramic photography by Douglas Cape, extends mechanisms featured in Cayley's 'prehistoric' title *The Speaking Clock*, inscribing generative characteristics into temporal elements extracted from another collaborative online work, an interactive dramatic narrative titled *What We Will* (2001–4).[62] In *What We Will*, viewers track movements of characters through London using an interactive mechanism built into the interface.[63] *Wotclock*, like *What We Will*, runs as a collection of QuickTime movies encompassing visual panoramas, but with more controlled features. In *wotclock*, a lettrist 'clock', within which Cayley plots transitory stanzas, appears atop *What We Will*'s gradually alternating photographic panoramas. Within this overlaid visual framing clockwork, shifting lines of machine-modulated poetry provide *wotclock*'s linguistic focus and an axis of verbal reception. Significantly, Cayley controls interactivity in *wotclock* so that viewers can only manipulate perspectives of the panoramas, and must wait for time to pass in order for an alternative visual scenario and its corresponding poetry to appear.

Wotclock, in great contrast to *Mesostics for Dick Higgins* or *Arteroids*, transpires very slowly; new passages of text appear every five seconds. Like *The Speaking Clock*, *wotclock*'s poetry changes with the progression of the silent but visible digital clock built into the interface. *Wotclock's* algorithm, according to Cayley's website, 'selects words from underlying pieces of composed writing. The algorithm searches for words that contain the particular letters (=numbers) needed to tell the correct time', and the program here quasi-randomly chooses words that fit constraints from the given text.[64] Symbols on the clock's perimeter indicate the actual time of day, which viewers may decode. In Cayley's clock-themed works, the changes in language represent shifts in time. As such, the device creates not only a constantly changing poem but also one reflecting the unending passage of experience. Seeing the ongoing connection between the shifts in time and in the lines of the poem is an engaging if sometimes deliberately empty process.[65] Successive moments of disarticulation, passages that may not cohere, lead the ongoing narrative to fruitful realization, but in their protracted essence produce a type of vacant stimulation as viewers wait for expressive significance, for meaningful connections between language, image and time to emerge.

Two mechanisms reflect the actual time of day: a graduated timeline-style clock with roman numerals at the top, and the colour-marked letters appearing on the central clock-poem, which may be decoded by means of the arrangement on the clock's outer circle of ten letters (ET, E, T, A, N, I, O, S, L, R, E, EE) in the twelve positions that represent hours on a clock. The letters in these positions change from white to red in a clockwise movement every five seconds.[66] Words in the database used to generate the poems must correspond – by sharing at least one letter – with the letters needed to indicate time on the clock. This aspect of the work imposes a serious constraint on the language projected by the poem, which is already constrained by using given texts appropriated from *What We Will*. Many words with the letters E, T, A, N and I – the most commonly presented letters – exist, but the given text is not very large. Thus the vocabulary of the piece, while shifting over time, frequently recycles the same words, as seen in the repetitions of 'should', 'seeing', 'already', 'have' and 'here' in the example below (which transpires at 10:16–10:17 a.m.):

Already have
been down

here with
me before

here with
me you

been down
here should

have been
before past

hello who
have dearest

already have
really should

already have
already dearest

been with
heard this

be then
before seeing

heard I
seeing smthing

we will
been seeing

dearest I
really should

already have
we should.[67]

Syntax in the output is not necessarily coherent, e.g. 'hello who/have dearest//already have/really should//already have/already dearest', although Cayley's algorithm does select natural language collations which nonetheless allow the time to be correctly 'spelt'. As such, the poem has the effect, overall, of reflecting internal thought, which is not isolated by time or always lucid or rational. Fragments transmitted seem to rehearse something that is going to be said, based on something that has already happened. Repetitions become echoic and transmit understanding through their cyclical musicality. Passages connect words evoked by the poem's *personae*, revisiting past events, seemingly mulling over (composing) thoughts while moving in open spaces, '...taken/from now', as pronounced in Diag. 3.8:

Diag. 3.8. John Cayley. *Wotclock.*

Wotclock establishes topicality and pathos through its representation of human characters, mosaic (fragmentary) depiction of
experience, and reliance on memory. Given its existence as a highly
technologized project involving a group of characters in a global
network, discovering issues and modes of communication as a
predominant subject comes as no surprise. While the use of Short
Message Service (SMS) does not feature prominently in *wotclock*, it is
suggested in the title, 'wot' being an SMS abbreviation for 'what', and
in the language of verses created, as in 'smthing', above. In the scene
depicted in Diag. 3.8 (2:04 p.m.), and that occurring immediately
before it (which transpires in a café), mobile phones, representing
the contemporary mobility of communication, are among the iconography present. A mobile phone (containing an SMS) dominates the
appearance of other *wotclock* scenes (as seen in Diag. 3.9 below).

Beyond issues of communication, texts appearing on the screen
(within the clock) often evoke properties and processes of love, time,
place and making – themes that both stand alone and expressively
merge in the work. Location, longing, interpersonal connection, as
well as the general contemplative or pensive tone of the poem,
emerge in fragments such as:

me all
still whenever

she along
I wherever

be wanted
I hear

here have
might forgotten

needed wanted
still where,

and in this passage, from later in the day:

yet when
will share

time making
love when

here in
this city

the cannot
be too.[68]

The passage here alludes to eros, but the certitude of a physical connection is left in an uncertain state. In this example and in other sequences, depictions and utterances in *wotclock* not only address issues of communication, but make viewers aware that slight changes in perspective, time and scenic montage have meaningful resonances and add depth to this meditative work. At one juncture, the absence of a person who was once there, indicated by detritus left on a table, illustrates the passage of time.

Embedding alternating background images strengthens *wotclock*.

Every hour only two interactive panoramic visual scenarios accompany the clock as backdrop. The interactivity opportunity in *wotclock* in fact solves the one visual difficulty to this piece, which is reading white texts over poorly contrasting backgrounds (see 'wot' on lip, Diag. 3.9). By manually rotating the image, any erasure of content is reparable; viewers can re-position images to make texts legible. Presented in colour, primary images are paired with related, thematically supportive black and white pictures; both are strikingly vivid but each holds its own purpose. Cape's high-resolution colour pictures capture places and scenes; the greyscale images surrealistically contrast locations of action by inserting distorted, processed imagery melding physical (bodily) and structural architectures (buildings, objects) more overtly suggesting correspondence to abstract internalized narratives contained in the poetry. Connections between the two, even if not obvious, change the demeanour of the overall projection, making us aware that narrative emanating from the conjunction of elements resists superficiality.

Changes of visual states in *wotclock* correspond to the clock's ticking and the passage of time, but have the effect of surprising viewers because they occur without announcement or fanfare, akin to the sudden whistling of the bird on a cuckoo clock. In the sequence introduced below, the replacement (black and white) image appears as a pair of envelopes, onto which faces are superimposed. Text appearing over the course of this section reflects separation and longing, if not more:

the message
on fantasized (11:07)

been given
she have (11:11)

message me
here say (11:17)

me here
what fantasized (11:27).[69]

The pictographic and poetic combine to imply more than either yields alone.[70] In the subsequent scene, at first presenting itself as visually similar, the alternative image radically differs by showing a montage of distorted hands and the distorted interior of an ornate building, challenging viewers to consider new meanings for configurations of imagery contained within this work. Changes to physical landscape and the technological depiction of human activity evoke emotional instabilities. Viewers who make the effort of sustained reading have no choice but to note the subtlety and extremity of purpose stretching across the hours, ascribing thematic and dramatic significance to both. Changes in location (visual) and poem (textual), and connections between the two, are sometimes in sync with one another, following and suggesting a type of harmonious progression, but the sense of an alternative, interior monologue persists throughout.

Wotclock's clock is ever changing, but remarkably consistent, values which make it appealing. Nearly a year prior to the example above, I captured the following sequence during the same hour period:

message well
fantasized here

have fantasized
it will

well she
will have

me been
given she

fantasized message
it will

have me
will have.[71]

An enticing poetic ambiguity persists. Certain lines come close to repeating, and ultimately will repeat, but this characteristic is not so prominent as to distract viewers from an ongoing narrative. While the work might benefit if given a more extensive supply text, the minor shifts (permutations) in *wotclock* reverberate poetically and diffract the language of its underlying dialogues in stimulating ways.

Diag. 3.9. John Cayley. *Wotclock*.

The complexity and richness of Cayley's design evolves over many hours. To be absorbed and read properly, *wotclock* requires extended viewings, watching the drama linearly unfold over the course of a day. *Wotclock* will not please someone who wants to watch a quickly revealed poem; qualities embodied in the piece are protracted and gradual. Time-based works such as *wotclock* heighten our recognition of time passing at a constant speed and our lack of control over it. When we read and experience time passing, and when we concentrate on time and reading in themselves, without doing anything else, time's passage can seem very slow. And yet in these circumstances, thought and other internal processes cannot be shut down, and this is (at least) part of the meaning suggested by Cayley's poem. Viewers can drop in at any moment, and may be provoked or stimulated by what is seen and read. The likelihood that no one will spend an entire day watching the algorithm choosing words and spelling them into the clock is not an issue. That only part of the work will be seen because the many images embedded in *wotclock* are displayed over many hours is tacitly understood. The steady progress of this piece, compared to most other works, runs counter to characteristics typical of the WWW. While its imagery is vibrant, and its verbal information is constantly generated, both its visual and verbal aspects have the qualities of musical drones, and the majority of content is withheld: inaccessible at any given time, *until its time comes*. Like many digital poems, *wotclock* does not insist upon a complete reading, but its slow pace is truly unusual. Cayley has produced an enigmatic but carefully elaborated and elegant work incorporating media and interactivity, one that does not strive to enthral by way of spectacle or by deploying many of the other popular commercial methods or strategies that we see in other digital poetry.

To mimic the types of phrases created by the poem, *wotclock* is: 'smething which/in time//makes smething'. Viewers familiar with the flexibility of navigation in *What We Will* will be shocked and possibly displeased at having to wait for whatever narrative the clock retains. A patience-testing work, *wotclock* is perhaps the only gradually evolving digital poem that transpires over a period of several hours.[72] Recalling Cayley's activities as a sometime translator of Buddhist texts, a reader might begin to consider *wotclock* as the translation of routines of existence and introspection, however contrived or artificially constructed, into art. From one perspective, *wotclock*, by cutting up text from another source and using a temporal formula to reconfigure it, may be seen as fusing Dada and Oulipo. Many authors before Cayley have sampled the language of one text to produce another; it is, however, the *quasi-* aspect of the randomness that makes this work both variable and poetic. Cayley folds in language to create poetry based literally on the time of day. Time mechanically (numerically) influences the creation of the poetic line and is simultaneously – if cryptically – projected onto the interface itself. What other poem ever managed to inform readers of the time, in addition to presenting contemplative verses?

Deena Larsen

Carving in Possibilities (2001)[73]

Deena Larsen began working as an electronic writer in the 1980s, and has described her 'addiction' to hypertext as 'probably-incurable'.[74] During the past decade she has produced numerous works in a range of styles on the WWW; the categories of links to her site reflect her artistic versatility: 'Intertwined tales', 'Flash collaborations', 'Matrix poem', 'Flash imagery work', 'Mystery novel', 'Kanji-kus' (short poems) and 'Structural work'.[75]

In this case study, I want to look at Larsen's *Carving in Possibilities*. Larsen stages both segments of this work in conjunction with a singular image: the face of Michelangelo's 'David'. Upon mouseover, poetic phrases momentarily emerge atop the picture, one at a time (i.e. two lines never materialize at the same time), remaining on the screen for as long as the viewer holds the mouse in place. Blurred and nondescript initially, the background picture comes into focus as verbal content is transmitted. As readers absorb language, an overall message begins to emerge and the shifting background of David's face becomes vivid. As such, the status of the image mirrors the level of understanding available to viewers of Larsen's poem. Reading and understanding happen incrementally.

From the beginning, Larsen foreshadows what readers can expect. The work's title page features a simple instruction: 'Mouse slowly to carve out your existence', and the start button activates the suggestive message 'and remember where you put your ghosts'.[76] In this manner Larsen suggests two contexts for understanding the work. First, viewers bear active responsibility in the poem – i.e. viewers make something through their responses to the work. Secondly, Larsen foregrounds memory's importance within the work. Another element that prepares readers is the short soundtrack that plays briefly at the outset of the piece. This music creates a fearful mood, dark in tone and eerie, which leads viewers to expect haunted and haunting themes. Two other elements extend the work's ominous properties. As mouseovers occur, sounds resembling muffled gunshots, reminiscent of those heard in video games, interrupt the soundtrack. The silence that follows the soundtrack is also foreboding.

Larsen embeds *Carving in Possibilities*' lines strategically. For example, she organizes links in a way that introduces overall thematic content by placing specific lines in the same area of the screen as the start button, which every viewer must activate. Several passages related to 'stone' are plotted in this section, such as: 'others had tried the stone/before me', and 'I saw precisely/what the stone was/meant to be'; nearby, the lines 'It's not a question of living/ in the stone' and 'but of living with the stone' are clustered.[77] In this manner Larsen foregrounds and acknowledges the lineage of historic artistry and thereby, by default of media in play, invites rumination on the implications of not working with something lasting and solid. Other lines thematically important to the poem located here include those containing historical reference ('I wanted all of you to guide/ my Florence') and authorial proposition ('I do not know my dream/in your reality').[78] At this juncture, the statue seems to be speaking, but this is not true in every passage. Further study reveals a multitude of voices and perspectives, as some lines describe the pain and agony the sculpture might have endured or would experience if it contained spirit.

In strategic gatherings of phrases – as in the proximity of the lines 'Which ghosts are you hiding from' to 'under the shelter of possibility' elsewhere – Larsen manages to strengthen themes and concepts within the nonlinear construct.[79] Another clear example of this occurs when, upon the imposed completion of a reading, she gives viewers the option to 'sculpt again', and proceeds to initiate 'Part II' by prominently placing the line 'Do you care who is/ telling you this now?' at its outset.[80] This tactic essentially allows for the introduction of new narrative dimensions and threads (e.g. 'A baby cried in the hot, still hot air') resulting from the placement of particular lines directly beneath the 'sculpt again' button.[81] Verbal, mapped content is the same in 'Part II' but viewers have a new launch point and consequently confront a different sequence of poetic materials.

Despite the numerous lyrical strands presented, a sense of address emerges, sometimes marked by use of phrasings that employ 'you' or 'your'. Within the poetic abstraction, a lingering diction enables flow and invites those encountering *Carving in Possibilities* to contemplate not only the stories Larsen may be telling pertaining to the statue but their own position with regard to a piece of flexible digital art. In the following sequence of transcribed passages, the nonlinear discursiveness intelligibly connects; '//' indicates a new link:

of your reality//
imperfections//
to see you as God//
Which ghosts are you hiding
from?//
Who breathed life into whom?//
indivisible//
I cut away everything
that wasn't needed.//
Is it the stone
or is it the name
that gives life?//
It's the same stone.//
no matter who inhabits it//
Is this what you want
to call back as real?//
The air was hot that day
I stood still
on dead grasses.[82]

In this excerpt, Larsen manages to compose and engineer phrases that effectively open into each other. It contains references to the act of sculpting in the elegant (unpremeditated) pairing of 'indivisible//I cut away everything/that wasn't needed'; thus the poem's waveringly embedded themes (e.g. stone, ghosts) again emerge. Ongoing thematic ties persist between stone and ghosts, elements viewers may connect with by associating stone to body, and ghost to soul. By injecting seemingly unrelated motifs (e.g. 'The air was hot that day/I stood still/on dead grasses'), Larsen achieves a type of capaciousness. *Carving in Possibilities* can be read as a speculation on the 'life' of a statue (or a stone, *any* stone) and the events it experiences; it considers whether or not such stones have a soul. For example, the line, 'Does your David lie only on the surface of your polished stone?' questions whether or not a statue, if doctored or worn, retains its soul. Other lines propose the statue as an empty 'being', and that a rock's surface is its core. Other connotations pertaining to characteristics of humans (perhaps even Michelangelo directly) may be read into many of the work's lines. *Carving in Possibilities* provokes contemplation of the idea that we are all stones who in a sense find our true 'image' through life experiences and how we individually control and negotiate what crosses our paths.

As an artifact created by a skilled hypertext author, *Carving in Possibilities* represents a significant shift in linking strategies; its mechanics reflect one of the alternative directions taken in hypertext practice since the WWW began. Specifically, Larsen's approach enables connections between links without the need to produce multiple files. The overall visual scenario incrementally shifts as texts rapidly appear and disappear, all of which occurs instantaneously on the screen without opening any new files. Flash enables Larsen to create a hypertext poem that would be difficult to create by other means. The nonlinearity she forces on readers partially results from the lack of a 'meta' view of the poem, an attribute often used by hypertext authors in the past that is not as much privileged on the WWW.[83] Without being offered a sense of the scale or location of content, viewers encounter a poem whose words are cloaked beneath the surface, whose appearance depends on motion and interaction and cannot be navigated linearly. By design, every single reading of *Carving in Possibilities* will contain similarities but ultimately differ. By virtue of this attribute, the work resides within a lineage still under development: Larsen demonstrates hypertext's branching out into new areas.

Carving in Possibilities features conditions of self-imposed boundaries resulting from the scale of its content. Unless readers make an effort not to do so, coming to the same point on the screen is inevitable. Some viewers will find this repetition tedious while others will find it echoic. To minimize this aspect of the work – which is not necessarily negative, especially since these variations attest to a certain level of poetic versatility – an author could further limit this constraint through programming (i.e. by coding each mouseover text to appear only once), or by adding more elements (i.e. words, phrases) on a more intricately plotted grid and thereby making the work more expansive.

Another aspect of the work not relayed through transcriptions such as those above is its game-like quality. While no score is kept, viewers progress and acquire meaning from the structure via movements accentuated by sounds. This comparison breaks down, however, in that one series of movements cannot be perceived as being superior to another. While variety in range of movement enables a more diversified poem, a viewing (or comparison of texts produced) cannot be easily qualified on the basis of the paths taken to a definitive endpoint.

Larsen overtly connects physical and digital artforms: while constructing the latter she provides viewers with an opportunity to command, shape and mould unique and accessible poems. Physical aspects of the work unquestionably exist. In defining the poem's scope, Larsen posits that the user 'carves' the face 'out of speculations about David, the crowd watching David and Goliath, the sculptor, and the crowds viewing the sculpture'.[84] Use of carving as a metaphor is curious, however, even if purposeful for Larsen's topical focus. In reality, viewers trace or track locations on the screen. Viewers apply labour to bring the text and image on the screen into focus, but they cannot make a wrong move that would destroy David's face, as could happen when literally carving a stone. Ultimately, progression of the onscreen (background) image is the same for every viewer – an impossible uniformity in the realm of the act of sculpting – though certain links, i.e. links on certain areas of the screen, alter the image more quickly.[85] Changes in the picture are, in every instance, linear, and viewers control the rate of emergence. As more text appears – as the poem accumulates verbally – the sculpture proceeds from an occluded state (Diag. 3.10) to one of bright visibility (Diag. 3.11). The face is already largely formed at the outset; viewers do not literally whittle away at a virtual block of stone. Larsen's distortions of the original photograph blur the image but the extant texture does not impede recognition; she provides an unambiguous, nearly completed artistic structure that becomes clarified as viewers absorb texts along unique, self-guided pathways.

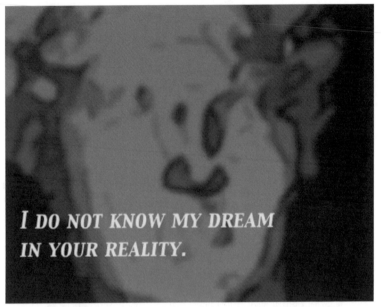

Diag. 3.10. Deena Larsen. Early stage, *Carving in Possibilities*.

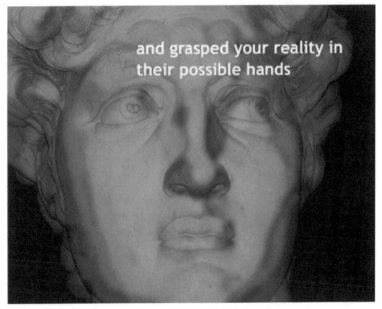

Diag. 3.11. Deena Larsen. Late stage, *Carving in Possibilities*.

Larsen uses fundamental artistic devices. She employs at least twelve different typefaces (fonts) and colour combinations, which may or may not represent different voices or threads in the work. Coloration and shape of the letters theoretically offer viewers a structure around which to read the poem. For example, in the dozens of 'hotspots' that activate verbal content on the screen, red text appears exactly once: when the viewer moves the mouse over David's lips and the phrase 'Be honest' boldly appears.[86] Given the placement of the link, the phrase non-arbitrarily conveys a message central to the work. Larsen presents most of her texts on a similar scale (approximately 18–24 point font), so on the few occasions where enlarged or reduced sized phrases appear, distortion of the typeface as such serves to transmit importance or emphasis. In one location at the top of the image, when the very small phrase 'or was there, always and ever/only one David for you' appears, the poem's voice seemingly softens to a whisper.[87] This type of address and modulation in the poem adds a layer of meaning; referring back to the sculpture, Larsen's ominous phrases (and inferred symbolism) expand the narrative.

Each interaction with *Carving in Possibilities* follows a different trajectory, although certain messages are pervasive and unavoidable. Many of the lines raise questions and doubts, which transmit a sense of yearning, if not soul searching: is the act of sculpting (or carving) pursued for pleasure, or does it represent a desire for something more? Are Larsen's questions rhetorical? We cannot suppose so; engaged viewers should find themselves searching for their own responses while carving out their existence. Every narrative begins at one of the lines near the bottom right (first reading) or left (subsequent readings), but from those starting points viewers can move anywhere and would end at the same place every time only if they intended to do so. The speed of movement through the poem and the methods for activating rollover links are entirely viewer directed, yet certain themes are omnipresent so reading *every* line veiled beneath the image is not required. Gradually taking in the evolving, accumulating, projected content in this piece does not present an extraordinary burden. In other words, audiences can unhurriedly absorb and react to the content. *Carving in Possibilities* is transitory; by design, it vanishes as it progresses, only leaving behind the trace that registers in the viewer's consciousness. Everyone records the work's messages in different patterns, yet common understandings of the work may be reached.

Carving in Possibilities is, in essence, a poet's digital poem: its lines are composed to make a specific piece of art through the contours of another piece of art. Lines embedded in Larsen's virtual structure emanate from her ruminations on a subject or theme; these are presented for the interactive viewer to order and ponder. The structure she employs has the effect of cloaking her lines of poetry, some of which viewers will retrieve. *Carving in Possibilities* gives its audience fragments organized around thoughtful, foreshadowed themes that engage each individual's imagination. Ultimately, the content of *Carving in Possibilities* is determined by a viewer's unguided contact with verbal, visual and sonic expression strategically devised by an author; as such, it is unquestionably a product of the digital age.

Jim Rosenberg

Diagrams Series 6: 6.4 (2005)[88]

Jim Rosenberg, a professional Information Technology specialist and experimental poet, is one of the pioneers and a true innovator in the field of digital poetry. Long before the WWW's emergence, Rosenberg wrote programs, created interactive poems and theorized about the mechanics of digital textuality, with a particular focus on hypertext, largely promoting alternatives to the archetypal link-node model.[89] He has been creating Diagram poems, 'the absolute core' of his work, since the 1960s.[90]

As discussed in *Prehistoric Digital Poetry*, Rosenberg employs a series of specialized processes to create the phrases that appear in his participatory diagrammatic poems. He prepares these poems in interconnected, palimpsest layers without using images.[91] Sections of his work present unobstructed words, as well as words and phrases visually layered atop one another. He introduces unique and radical hypertext syntax and conceptualizations for readers to consider using in their negotiations with the diagrams. The sophistication of these features indicates the enormous intellectual properties of his efforts.[92] For example, Rosenberg refers to his atypical authorial approach as 'conjunctive' hypertext. Pursuing this approach, writes Rosenberg, 'presents activities as elements to be *combined* into a whole effect'.[93] Instead of using the common link-node model comprised of a series of links, Rosenberg – like Larsen – layers his poems so that the activation of different layers depends on the cursor's screen placement. However, unlike Larsen, all Rosenberg's verbal components are always present. This technical and aesthetic (visual) structure built by Rosenberg is a convenient vehicle for the conjunctive aspect of the work because it creates hypertext activity for the reader on every screen, even when the reader does not follow links. 'Conjunctive hypertext', writes Rosenberg, 'refers to constructions where the relationship between a component and its elements is 'and' rather than 'or'.[94] Readers begin to see the screen texts as pieces of a puzzle. Instead of changing focus and paths while traversing the poem, the reader experiences conceptualized, co-existing, co-operative advancement in the conjunction of all poetic passages contained in the diagram. 'It all interacts', concludes Rosenberg in 'The Interactive Diagram Sentence.[95] Rosenberg presents his Diagrams as simultaneities, as textual events that happen all at once. They demand that readers absorb a lot of abstract information, insist that intertextual and personal connections or associations are made, and associate words and written messages together in entirely new ways.

Prehistoric Digital Poetry refers to the diagram notations devised by Rosenberg – which direct 'verb' and 'modifier' activity in the poems – as a means by which to navigate through the work. On a simple register this is accurate, but the function of these notations is more profound: Rosenberg designs his syntax as an active syntax, or, as Rosenberg writes, an 'external' syntax, 'allowing word objects to carry interactivity deep inside the sentence' and not merely as a navigational device.[96] This approach, as Hélène Perrin and Arnaud Regnauld write in the essay 'Jim Rosenberg's *Diagrams Poems Series #3*', achieves the effect of 'hypertextualizing syntax', which creates an effect where hypertext 'simultaneously unravels and complexifies the structure of language'.[97] Readers use embedded notation symbols to construct an understanding of the intentions of Rosenberg's virtual object. Diagram notation literally 'gives the 'parts of speech' that get disrupted when words are put on top of one another. The layers are pieces of a cluster that happens all at once.'[98] In comparison to the relatively basic (if thoughtful) work prepared by Larsen in *Carving in Possibilities*, reading poems devised using such a scheme becomes an enormous cognitive challenge, in part because verbal fusions created by the notation are not always clear, but also because, as Perrin and Regnault elegantly write, 'the externalization of syntax and the transformation of hyperlinks into nodes constitutes an interrogation of the relational as a form of (Heraclitean) continuity which rejects the binarism traditionally associated with the herme-neutics of the sign as form of meaning.'[99] Many fragments must be simultaneously absorbed, considered, recalled and conjoined, without the expectation of establishing linear intent.

In 2003 Rosenberg discovered a program called Squeak[100] that permits online replication of his inventive structures, and two complete series of his Diagram poems have since been launched on his website.[101] Following Rosenberg's poetic construct, readers can begin their immersion anywhere. Entry-points provide non-verbal overviews of each Diagram contained within a title; thumbnail depictions of each simultaneity link to the poetry. The example I have selected to explore, Diagram 6.4 (Diag. 3.12), contains four word 'complexes'; three are connected by verb symbols, ringing a central complex:

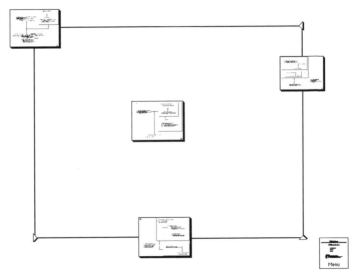

Diag. 3.12. Jim Rosenberg. Diagram Series 6.4.

In this slightly asymmetrical diagram (from a diagonal perspective), an obvious disconnection exists between the central complex and those around it. This disconnect seems to indicate two separate but related threads that readers might consciously connect in the poem. Observing this dynamic – a node within a verbal loop of nodes – would seem important while reading the work, yet where the exploration starts is left open. Normally, this decision might heavily influence a reader's understanding of a hypertext poem, but, as revealed in the discussion below, the starting point is not a major factor in Rosenberg's Diagram poems.

Given the complexity and challenges of Rosenberg's schematic designs, building a relationship and understanding of how the three interconnected complexes work together is an immense task since various relationships are put forth. The discussion below – an example of a plausible complete reading of one simultaneity – attempts to provide a reasonable assessment of what transpires and how a literary audience might appreciate the work; this is followed by a summary of observations regarding other sections of the poem.

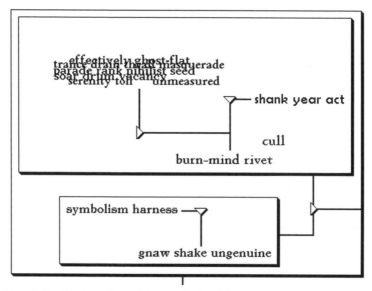

Diag. 3.13. Jim Rosenberg. Diagram Series 6.4.

Linking to a simultaneity to interact with the various elements (sometimes including nested simultaneities), readers face a meta-view of a section of the poem. The section represented in Diag. 3.13 is the complex on the upper right side of Diagram 6.4, which contains three distinct components: two clusters, at right and bottom, and an inner-complex (a rectangle comprised of a pair of smaller rectangles, one of which contains a cluster at upper-left).

Looking at the verb cluster at right, the reader sees, before any interaction, some or all of the following words and phrases (or even more, depending on how closely s/he looks): 'too shudder charnel', 'ballast grinder's', 'chant mind arrow', 'delirious', 'duneless'.[102] At a glance nothing particularly unifies these – although readers who seek to build a conjunctive foundation upon which to understand the poem may identify the theme as something 'worn down': this is suggested by grinder's, duneless, charnel. Opening the cluster, three separate passages (slots) appear within the overall simultaneity: 'too shudder charnel/chant mind arrow/duneless', 'ballast grinder's/facial sheath stump', and 'kite force body/weave stream flock delirious'.[103] Despite a pleasant internal and probably non-accidental lyric that happens when readers encounter poetic information in certain orders, e.g. effectively pairing 'duneless' and 'delirious', the passages themselves do not illuminate any further thematic quality. The lines 'chant mind arrow' and 'kite force body/weave stream flock delirious' may offer an antidote, escape from or resistance to an otherwise dreary sense. Since this cluster is a verb, it might be considered to be something wearing down but fighting its conditions as such.

According to the design of Rosenberg's notation system, the 'result' for the reader of the cluster on the right serves as the verb for what the reader discovers and comprehends about the bottom cluster and the accumulation presented by the rectangles above it. Modifying symbols are not presented, thus the language is active, perhaps beyond the point of precise determinacy. The poem progresses in any number of possible ways, depending on what the reader absorbs and makes sense of. An issue inevitably arising out of these atypical verbal configurations is the matter of address in the poem. Since pronouns or finite subjects do not appear, may it be assumed that the reader, as a member of technologized culture, i.e. someone who is reading computer-based literature, is being directly addressed? Presuming a type of impersonal but objective communication seems most reasonable since readers receive the poem's messages and respond individually to its dynamics and content. As in any poem, a one-to-one relationship between reader and poetic mechanism transpires despite the unconventional presentation. Decisions, or contexts created by the reader, regarding the role of the two clusters dictate an individual understanding of the composition, although the parts of speech in the diagram, apart from the cluster identified by the verb symbol, are undefined.

Although any meaning inscribed in and taken away from the work may be utterly ambiguous, tangible declarations can be wrought through an interpretation of Rosenberg's conjunctive practice. The reader may identify the cluster at the bottom as a noun; registering readable words in the cluster before interaction occurs – 'soliloquy dam sparrow,' 'hinge,' 'mercy grommet' – seems to confirm this presumption. Uncertainty persists, however. As Perrin and Regnault observe of Rosenberg's work, 'words within the clusters have been stripped bare of any former usage value and grammatical function', and instead 'coalesce into semantic haloes brimming with the minute molecular interplay of semes and phonemes'.[104] Slots here include the passages 'soliloquy dam sparrow/mercy grommet' and 'crosswind declown marrow'.[105] Rosenberg's verbal imagery is stark and vivid, yet mysterious as well. What is a 'mercy grommet'? A grommet made with mercy or something that drives a hole through mercy? How does this relate to 'soliloquy dam sparrow', an image anyone can conjure without much difficulty? In 'crosswind declown marrow' readers find a rhyme and might connect it with the line about the sparrow. In the third layer, however, a further circumstance arises when the reader encounters a nested cluster within the slot that opens with the clear message 'extremity hinge'.[106] These lines, essentially obstructed from view otherwise, suggest a stormy passage emerging, e.g. 'miscue gale hole', 'ballad skin triage hinge'. What seems to be indicated as subject matter is a tenuous, if not tumultuous, predicament in the natural world.

Readers following the above scenario can make connections between subject and verb clusters but then must interpret and identify the upper structure's 'message' before proceeding to plug this complex into the larger diagrammatic equation. Alternative readings and reading methods are, of course, also possible. To fully experience this section of the poem, we must find a way to make sense of the textual structure shown at the upper left of Diag. 3.13. Visually, this conglomeration occupies most of the screen due to its composition (the structure of the diagram), although Rosenberg offers no indication or reason to believe it carries greater importance due to its scale. Shape and activity of the structure do, however, place increased demands on the reader. Here, the role of the combination of rectangles is difficult to ascertain. Readers might see it as the object in a subject-verb-object structure, but we cannot presume this with confidence because it also appears that the two external clusters may serve as a verb – the inner rectangles connect to the outer rectangle via verb symbol. Alternatively, the metastructure may be subject-verb-verb, or from another point of view the two clusters' structures may be 'verbing' (i.e. connect as a verb to) the object. If the lower cluster is a subject, can two actions be attached to it, or does the result of the subject-verb cluster then become a verb of some sort? The fact that the external syntax is not always straightforward adds complexity to the task of interpretation.[107]

The lower internal rectangle in the upper section is perhaps the easiest and most straightforward of any we have yet encountered: 'symbolism harness [connects by verb symbol to] gnaw shake ungenuine.'[108] The latter line perfectly indicates Rosenberg's authorial style, wherein a verb is not portrayed as a singular action but as a nuanced sketch; in other words he adorns the action and vivifies it with character, i.e. 'ungenuine', which here somehow questions the character of the event. Despite the directness, however, he challenges readers by establishing a predicament where they must develop a cognitive association between the passage (which seems to prescribe casting off false restrictions) and what appears above (discussed below) and then determine how to connect them with the external clusters. Since all this occurs in the process of reading just one section, comparable to a stanza or verse of a larger work, the magnitude of demand, i.e. the effort required to receive them fully, becomes undeniable.

Directly above, 'cull/burn-mind rivet' appears to be the verb, or verb clause, between both the cluster and the curious, nonfigurative, stand-alone phrase 'shank year act'.[109] Depending on how the audience 'sees' the diagram (Diag. 3.13), however, it is also possible that a one-to-one relationship exists between 'burn-mind rivet' and 'shank year act', and between 'cull' and the cluster. Viewing the construct in this manner perhaps simplifies the making of meaning, but multiple possibilities exist nonetheless. The conjunctive mode, as Perrin and Regnault observe, 'suggests rather than defines'; a meaning 'is not embedded in words but results from action and relation'.[110] Nonetheless, 'burn-mind rivet', combined with 'shank year act', might be understood as finding and using (metaphorically) something that pierces one's mind (recalling the function of 'grommet', above). How the combination of fragments connects with the other internal word combinations, i.e. the cluster and 'cull', then within the overall structure, is not entirely clear, yet establishing such a relationship – somehow, and perhaps not definitively – is the reader's work, given the conjunctive context.

If a reader hopes to distil something meaningful from the content, the best way to describe the experience of interpretation is that the reader is required to unfold a series of enigmatic clauses to reach a gradual understanding. With 'cull' as the verb, an indication to conjure or locate the images unveiled in the cluster presents itself as a possibility. In the cluster, the single word 'unmeasured' stands alone until the reader triggers the following passages: 'trance drain thrall masquerade/serenity toll unmeasured', 'effectively ghost-flat/ soar drum vacancy' and 'parade rank nihilist seed'.[111] How 'cull' relates to these phrases, which in some senses reiterate bleak thematic tones found previously, in a similar abstract yet expansive style, is another matter. Are they the issues, poetically stated, that the poem culls? Are they what the reader must cull to imagine the poem? Questions arise, such as what is a 'trance drain thrall' and how is it a masquerade? What is 'effectively ghost-flat' or 'soar drum vacancy'? How do we cull or why is the poem culling these images as one? What exactly is the composite relation to the 'shank year act'/'burn-mind rivet'? The poem can be read as an observation and as a challenge to or critique of culture, perhaps indicated by 'parade rank nihilist seed'; a reader might make some sense of these lines by connoting that our civilization – and this could mean Western, North American, or some other – lacks spirit, that an opening to transcend this condition, i.e. the 'trance drain thrall', through percussion or music, and possibly inventive communication, is available. Integration of the line 'serenity toll unmeasured' supports the idea that contemplative strategies offer a possible solution to the problems the poems seem to address. How the sensibility derived from the upper portion conjuncts with 'symbolism harness/gnaw shake ungenuine' is not explicit, but if the poem is regarded as a vehicle for transformation, the information could be synthesized to form the sense that a reader (or, by extension, the culture) needs to find ways to use the ideas presented in utterances to invoke (impose) changes.

To synthesize all of Rosenberg's poetic notations and projections in this section alone, to create a unified node of expression within just a part of the overall diagram, is an unfamiliar activity but not impossible. Descriptions of the text and process of reading are not comparable to experiencing the poetry first-hand and do not justify the ambitious intent of the conjunctive practice, yet are necessary if we are to follow through with a reading of this portion of Diagram 6.4. A reader must find ways to both accumulate and distil sensibilities within a construct meant to be multiple and textually expansive. Fortunately, in this example, conceptually fusing the elements is facilitated by commonalities found within its parts. The collected verse shares a perceivable direction. All of the elements together are what Rosenberg's spatial, interactive poem is: a call to thought and a resistance to ordinary forms and the status quo in expression, i.e. complacency in writing, reading. By combining all the layered data and verbal notation, readers can detect a need for recognition of circumstance as well as the need to address and act on it. Picking up and reassembling the pieces of the poem, the audience receives a fragment of the whole, a serious foundation from which to inspect and comprehend the other simultaneities contained in the series.

What happens when another of the ring complexes of Diagram 6.4 is explored? Inside the complex at the bottom (Diag. 3.12), joined to the previously simultaneity by verb symbol (i.e. the first complex – discussed above – acts as verb to the second), a cluster serves as the verb to a pair of rectangles, both of which combine syntactically and connect stand-alone passages with clusters; the upper cluster itself also contains an internal diagram sentence. The compound writing style found in these unobstructed passages are easy to associate with lines explored above and contain identical themes and approaches to presenting poetic information. For example, the first line of the now familiar sounding sequence 'mirage storm archetype friction/crown fake rightness/thunder sequence', implies an unreal (or empty) sense of what is said in the line (if not the entire passage) in the same way 'ungenuine' and 'masquerade' do (above).[112] The other clearly readable passage, identified as the modifier of the small cluster appearing above it, 'careen transcend-ghost/backstay windup', re-inscribes inanimate spirit into the work, as does the opening screen of the upper-left simultaneity of Diagram 6.4 ('trickster crouch', 'ghost-song/mime roll melter' or 'star sort shirk').[113] Each passage adds one of multiple dimensions to the overall complex and diagram, and identifiable threads emerge; the accumulation of lines does not feel like a list but rather a unified yet variable litany. Readers sometimes face, as above, clusters embedded within clusters. In the visual hierarchies that result from Rosenberg's design and process, the last layer activated becomes a momentarily prioritized word (concept) in the construction; this happens, for example, with 'nil mouth siren' and other words and phrases in the upper left simultaneity of Diagram 6.4. Rosenberg creates a polylinear situation throughout, in which the reader has no choice but to see this text in the ultimate slot alongside 'trickster crouch' etc., an arrangement that can shift with the slightest action.

At this juncture, when looking at similarities in style, theme and poetic approach in the verb(al)-ring of Diagram 6.4, the content of the centre complex becomes key. Certainly, despite the lack of connection through diagrammatic notation, a relationship must exist between each aspect of the Diagram, especially given Rosenberg's conjunctive sense of poetic (digital) architecture. Opening the centre simultaneity, this presumption is immediately confirmed when the unobstructed phrase 'wholeness cobble cull/weep too lateral' appears (among others). As above, 'cull' figures into the simultaneity in a significant verbal role, as if this is what the poem does or asks the reader to do (or make) of the work. Since the passage is so prominently displayed, readers cannot help but see it peripherally at the same time as they explore the three clusters that appear concur-rently. The line is a serious charge toward pursuing 'wholeness' and involves the recognition that crying is too flat, or that release and stasis are part of the textual equation. Diagram 6.4, if it is about anything, seems to be about the pursuit of discovery, ordered and undertaken by the poet, who urges readers to do the same. Further exploring this simultaneity, additional lines, such as 'world tongue density/collide-scree eyes', offer clues as to the intentions and directions of the poem, and aesthetically resonate with other passages contained in the Diagram.[114] Internal music and lyricism occasionally emerge, as above, in cluster juxtapositions such as 'refold scar undertow' and 'life escrow goad mark'. Although they are diagrammatically disconnected, the simultaneities in Diagram 6.4 are comparable to one another. The outer ring triples and activates motifs reflected in the nucleus of the Diagram. If readers were to enter the central complex first, followed by those that surround it, they might in fact recognize a centrifugal effect, where themes established in the stand-alone complex (the core) become further refined as they are doubly (or triply or quadruply) reflected in the exterior complexes.

Recognizing Rosenberg's conception of hypertext 'as a medium thought', rather than simply as a type of link-node structure, is extremely relevant to understanding his work. Throughout the Diagram Series, it is important to realize that the hypertext design does not impart one-to-one associations between disparate documents, but rather serves, in conjunction with the externalization of syntax, 'to carry the infrastructures of language itself'.[115] Since the platform of expression is computer-based, the reader experiences grammar and language in an unprecedented way, particularly given the synaptic connotation suggested by Rosenberg's concept. He writes, 'To understand the network one *becomes* the network.'[116] An alternative idea I would like to offer, paraphrasing Rosenberg, is that to understand the poem the reader must *become* the poem. Mechanics within the poetry mirror the processes of thinking, where a reader, as Perrin and Regnault write, is 'constantly making, breaking, losing and simultaneously redistributing multiple connections'.[117] These conditions foist exorbitant demands on anyone exploring the work because the poems are multiple. At the same time, they are emitted from a single place and given uniform typography and colour, so they also appear to be univocal. Organized by the poet but unique to the reader, the poem happens in the accumulation of the read (and viewer-processed) grouped lines and phrases, and in the cross-identifications, understandings and meta-connections proposed by

Rosenberg's conjunctive scheme. His aspirations can be likened to creating chords with phrases instead of notes with words. Further, his notation indicates we must read (or hear) these chords *within* the context of other chords (or fused structures) to experience the poetry fully. Essentially, as readers, we must consider each of his diagrams as a whole unit, tuned together. As Perrin and Regnauld observe when translating Rosenberg's poems, 'Words resonate within clusters, reverberate several meanings, reflect etymology, echo each other across clusters and across diagrams, across the whole series, and finally carry with them the ghost of the old context they have left behind.'[118] These unusual expectations for the reader are overwhelming, yet also refreshing (if not downright exciting) in terms of poetic innovation.

Synthesizing so many elements at once is unquestionably unfamiliar reading territory. Through his use of technology, the number of dimensions reached in Rosenberg's layered alphanumeric structures exceeds the expressive boundaries of even the most intricately threaded and complex written (multi-vocal) experiments in poetry. In fact, absorbing them as they are meant to be absorbed is practically beyond our capability in time. At very least, engaging in the Diagram poems is extraordinarily demanding; a reader must understand and follow the schematics and also develop the ability to retain and make sense of the layers and fragmented passages.

Rosenberg addressed the problem of, and need for, increased – yet unified – variables in the commentary included with 'Diagrams Series 6':

> … we must all occupy the same world space, do no harm, and yet be free. Carrying multiplicity inside the thought, inside the sentence: the thought as world. At a time when our world is in deep painful need of more multiplicity of thought.[119]

Are Rosenberg's poems the twenty-first century's version of crafted apocalyptic, anarchistic lyrics, created with contemporary tools? Do they intend to convey a particular message? While any inclination to compose with chordal language is impressive, it is typically difficult to interpret work that can be read in any number of ways, though not altogether impossible. Readers, with effort, may begin to absorb instinctively Rosenberg's system, and the poems he projects, as he wants them to be read. Employing typical strategies, readers will likely encounter difficulty. If the best they can do is patch together a poetic narrative by way of Rosenberg's dense and peculiar object, they will still achieve and experience something satisfactory, if mystifying. Since the Diagrams take time to grasp fully, and not everyone will make the effort to learn what conjunctive hypertext is, or how diagram notations function, Rosenberg's work invites different types of approaches. Some will not, unless guided otherwise, see his Diagrams as intended. While it is not possible to measure the profundity quotient of any given reading, or quantify degrees of reading, I would argue that any reader of the piece can build appreciation for the poem even with a basic interaction, or by hearing the author give a reading of the work, a scenario in which only one line is heard at a time.[120] While the diagrams clearly offer multiple possibilities, building an intuitive narrative of any sort without attending to the diagram notation may be just as viable a method of approaching these poems.

Synthetic operations in the process of composition, projected in multiple layers, create a textual environment in which the reader/participant must synthesize complexities of the object at once, as a single yet fragmented unit. What emerges, as with many digital poems, are flexible, noisy (if literally silent) poetic utterances freed of familiar controls and message delivery system. Again, the author structures words without being able to impose, finally, the order in which they will be received or how they will be read and understood. The author makes an effort to guide an encounter with the notation and explanation of the proposed functionality for the compiled elements. Meaning possibly emerges – depending on how much or even how little words resonate with the reader's imagination and experience, or what she or he can tune into. Even the most unusual and challenging digital poems have traditional roots. Rosenberg's processes and mode of presentation have not diverged much from where he began as a digital artist. As described in *Prehistoric Digital Poetry*, Rosenberg begins by writing words on a page. What he computationally does with the writing – with and without software – is, however, highly refined and unusual by any standard. Rosenberg invents new processes encompassing old and new methods; using

3000 our names smashed forgotten black dust everywhere it's
    gone
3000 untrue we're inside others just from the sheer mixing of the
3000 world september there were snows earlier than usual
    november
3000 oddly warm no fingers no memory.[132]

A 1996 calendar, the year of composition, and some lines about
that year appear before the section's concluding line: '0000 that
last tuesday screams last swallowed.'[133] Looking more closely at
this poem, this persona finds comfort in considering the future
from the present – or from a timeless place, perhaps – despite
his (and everything's) overt absence from it; ground zero, as
reflected in numbers, reflects turmoil. 'Blood.txt''s next section,
'III', seamlessly addresses the commercialism of the 'Net', water
and 'Net sex'. Poems by Sondheim appear next to poems by
others, such as Robert Creeley and Coleridge; some documents
directly address the ostensible subject (e.g. blood), while others
offer personal and philosophical observations on a stated subject
(e.g. 'Loss') that coincide with Sondheim's themes. As indicated
in Sondheim's explanation of process, though some of these
texts are computer generated, they mainly result from dialogues
and the author's musing in forms of writing that intrinsically
inscribe the words of others. Reading these works is sometimes
challenging due to lack of convention, but no more so than other
contemporary Modern and Postmodern peripatetic works. The
purpose and scale of *Internet Text* is grand, and not outside the
boundaries of consideration in the realm of digital poetry, but the
conditions of these texts do not adequately represent Sondheim's
contributions to the field, which he often achieves while working
with other artists. For this reason, I turn to focus on some of his
collaborations.

    Sondheim co-produced two distinct imagery-dominated collabo-
rations with Reiner Strasser, 'Dawn' and 'Tao'.[134] Whereas much of
the material in Sondheim's *Internet Text* consists of raw material
– either documentation, roughly mixed or written experimental
narratives, or standalone images – these works superimpose written
(and sonic) attributes with digitally morphing imagery. 'Dawn' and

'Tao' are similar in several ways. Though they are serious in tone and visually rich, in many regards they are elementary from a technological standpoint. Flash, programmed by Strasser, is used in both to combine text, imagery and sounds produced by Sondheim. Using software to synthesize disparate elements, a common practice during the WWW era, enables Sondheim and Strasser to propel a unified theme with stylistic refinements vastly superior to pre-WWW kinetic poems.

In 'Dawn', a work that features a linear continuum of pictures presented on a loop, Strasser blends high-resolution images, thereby endowing the screen with cinematic properties. Within four main projected images, which sequentially merge as the piece plays, Strasser skillfully embeds transitional, interstitial images that never quite fully appear but nevertheless correspond to the contours of the primary images and impart flow and elements of visual sublimity to the work. Eight separate textual passages, also looped, are inscribed upon these shifting images. These lines or stanzas only faintly overlap and no more than two appear at any one time (as seen in Diag. 3.14):

Diag. 3.14. Alan Sondheim and Reiner Strasser. 'Dawn.'

Combining the contrasts in colour between image and text with the appearance and fading of the text heightens the work's aesthetic properties and has a dramatic effect. The sequencing of texts unveils autobiographical discourse combined with reflections on a physical place and nature:

> my father lives alone in a large house in a small town
> my mother died five years ago
> at ninety-one he's moving slowly, he wouldn't
> be able to climb, here in the chorus of the dawn
> the early morning of the blue deer, the blue deer.[135]

As the images and text make their transition, a deeper, more painful philosophical dimension emerges: 'cursed be the god that presents so much death and beauty.'[136] Without the temporal component – in other words, if all the text were just presented at once as a whole – the shift in tone from one level of introspection to another would not be as striking. 'Dawn' is essentially a narrative poem featuring observations of self and reflective thoughts about the poet's father. Its synchronized unfurling of pathological sensation, however, creates a textual contour that subverts emotional flatness and leads viewers along in anxious wonder as they make connections between what is seen and said. Until the loop begins repeating, viewers are enticed to follow the text with a sense of suspense.[137] Although the pace of presentation of text is not rapid, the authors provide a (rollover) pause button – the only interactive feature of the title – that stops the text and images. Meanwhile the audio track continues, combining what sounds like the drone of insects with 'scratches' one might find at the end of a worn LP record. Together, the sounds, images and contemplative language immerse viewers into a human scenario, into a scene that could be our own but is not; we are effectively, if momentarily, transported into someone else's landscape and predicament, and stimulated by these circumstances. These are the factors that contribute to make 'Dawn' a success.

'Dawn' begins by celebrating life; the passage at the start of the loop is joyous:

> there are deer in the river, eleven blue herons

i am glad i am alive to see this
the singing and the morning of the world.[138]

Verses evolve to incorporate tones of lament, introspection and statements of need that by the end of the loop result in the discovery of wisdom: 'i will learn to listen/to the beginnings and endings of the world.'[139] Content portrayed in the images, at least two of which contain fences, relates to themes established by the text, which not only reflect the beauty of nature but also recognize our sometimes awkward place and the abstract things to be found within it. Fences imply human intrusion, perhaps confusion of order, and certainly division. Sondheim's words indicate an awakening in the author, juxtaposed with poignant images.

Loops are created in the form and content of the work; likewise, media elements are used to project the cyclical, triumphant powers of nature. Technology synthesizes these components, which is one of the most powerful compositional attributes of Flash. The design of 'Dawn' demands that text be read in conjunction with other media elements. The transitions between passages and the morphing of images are both gradual. In its simplicity, the looping imparts a contemplative dimension that helps viewers not feel rushed. The scenery of the poem is unchanging and can be read without much difficulty. As such, this is a gentle digital poem to follow and appreciate in a somewhat conventional sense. Instead of noise, the sounds and combination of other elements offer a type of tranquility.

Now let us turn to 'Tao', a short interactive piece, with minimal but consequential interactive properties. To describe this piece in outline, the viewer will see two mirror image videos, presented in different hues, that show travel footage shot from a moving vehicle. These videos can be 'flipped' (i.e. reversed) by the viewer. The videos appear above a text that gradually unveils itself over a span of 38 seconds. An abstract, digital object – some sort of unexplained avatar or apparition – is superimposed on the video segments. Additionally, the video on the right contains a layer representing a starry skyscape (see Diag. 3.15). This image may be seen as a parallel craft, a virtual sail, a ghost, or some other equally abstract entity, such as a galactic object 'blown down to earth'.[140] Technology renders supernatural animated imagery onto video, allows us to change the perspective of the visual components, and enables the authors to embed sound and animate text.

earth blown out to stars · stars blown down to earth by fast cars · baghdad and addresses of the invisible

Diag. 3.15. Alan Sondheim and Reiner Strasser. 'Tao.'

Although viewers are invited to interact with 'Tao', it can also run its short course without any manipulation; in other words its design never prevents the reception of text (all of which appears in Diag. 3.15). Words appear from left to right on the screen, as the soundtrack (Sondheim playing a shakuhachi flute) is amplified. Together, these elements create a sombre and wistful tone. The language of the poem amplifies this tone by invoking phenomena beyond the earthly plain and also by embodying a sense of frustration ostensibly caused by the war for oil, indicated by the reference to the capital of Iraq. Looking more closely at these words, they suggest that transcendence from the physical plain is blocked because even 'stars' are 'blown down to earth by fast cars'. Therefore the artists must implant the otherworldly into the poem through the abstract object that floats across the otherwise realistic, videographic presentation. Sondheim and Strasser imaginatively find ways to see and offer something 'invisible', that it is to say, something in addition to the typical content of mass media news, something beyond the earthly. Viewers can watch passively, but different meanings register when you begin to interact with the video. While the objective text cannot be changed, the alterable visual scenario, already kinetic, gives the viewer the ability, through effort, to create changes of perspective and distort the landscape witnessed. The abstract objects appear as if dancing with each other – not in perfect synchronization but based on the viewer's modulations. Viewers can transform both what is seen and what is done with what is seen. Through the 'addresses of the invisible', a hopeful message is delivered, one that suggests that effort will bring results, that alternative realities are attainable.[141] Perhaps there are such things as virtual spirits or perhaps this is a reflective escape; both invocations are plausible. Enigmatic manipulations of a videographic image combine with other elements to suggest possibilities: consider the ability we have to reverse perspective, or to build multiple combinations of perspectives. While not as simply devised as 'Dawn', 'Tao', as a result of its containment and scale, offers neither a terribly challenging nor an overwhelmingly mystifying experience. Ponderous information is, however, contained within the videographic and textual components; in this manner the authors have engineered an effective vehicle for poetic contemplation.

'The various issues of embodiment that will arrive with full-real VR', writes Sondheim in *Internet Text*, 'are already in embryonic existence.'[142] With this in mind, let us now focus on *Second Life*, an interactive entertainment environment created by Linden Lab in 2003, which Sondheim subverts for his own purposes. This online virtual world, which is free and available to the public, enormously amplifies the use of virtual figures by enabling its users to cultivate unique avatars on fanciful stages and operate in real time. To participate in *Second Life*, users obtain a 'client' (an application or system that accesses a server on the network, across the WWW) and software, and then make an avatar and create visual and/or interactive objects.[143] Several other features of *Second Life* make it unique: it is 3D, meaning that figures are driven backwards, forwards and in other directions, and anyone logged into the system can observe the movement of these figures.[144] Users build objects, constructed as parent and non-parent objects; scripts within objects activate textual and other occurrences, and can be used to create movement, or to interfere with it.[145] As in MOOs (before the WWW), creative programmers can make actions and events happen that could never occur in 'real life'. Unlike MOOs, *Second Life* experiences are much more visual, visceral and variable, as camera controls can be used to alter perspective(s) and manipulate angles of view.

In this expressive, experimental, virtual terrain Sondheim often collaborates with Sandy Baldwin.[146] Engagements by Sondheim and Baldwin show how literary texts, and the experience of literary activity, are transferred to a digital setting where imagined materials (including parallel bodies) and actions are visualized and produced spontaneously in congress with others. Baldwin has noted that Sondheim's new modalities of expression are often 'pre-symbolic' forms in which 'extreme … turns organic' and writing 'is not a sign but an organic membrane'.[147] What happens in these creative events, according to Baldwin, is a 'tethering of body to the screen'.[148] Avatars represent physicality, motion, and are capable of expressing or inspiring emotion.

Video (MP4 format) can be connected and played on the virtual objects encompassed in *Second Life*, so Sondheim and Baldwin have, among other experiments, created animations (videos) of avatars using motion capture technology. In addition to rendering lifelike movements, these avatars can be contorted into positions the human body normally cannot accomplish in the 'real' world. Practically anything that can be conceived can be visually achieved in *Second Life*, which gives its users unprecedented creative possibilities. Text (programmed instructions) can be scripted within objects and also manually entered in real time, similar to 'Chat' dialogue boxes. Sondheim and Baldwin use both approaches to present text, sometimes borrowing scripts from others and making alterations to them.[149]

Conspicuously, text is often absent in *Second Life*. However, Sondheim – who claims an inability to 'make distinctions between what's writing and what's not writing' – notes that he does not 'have to use words, because the space itself is its own text'.[150] The vivid imagery carries some of the poet's responsibilities, relieving him of some of the authorial duties. Working in *Second Life*, for Sondheim, adds a profound sense of wonder to what he can expressively accomplish; he has exploited its unusual capacities in several ways. Vertical movement, transporting avatars high off the ground, fascinates him the most, as it tends to make 'things disappear', with participants losing control over their surroundings. Sondheim uses this approach to 'try to get things that have no relations to things in the world', where a type of 'self-induction' occurs in which the avatars literally transcend ordinary circumstances.[151] Historically, writers have had to exercise a combination of imagination and language to reach such goals. However, someone with the skills to engineer objects in such a highly technologized environment as *Second Life* can fluidly combine objects and motifs to achieve the effect of transforming an unbelievable artifact into an authentic one.

Often *Second Life* gives one the sense of 'live' writing – instantaneous or circumstantial in the moment – through the use of scripts that activate texts and other visual occurrences. For example, when an avatar approaches another virtual object, scripts can trigger textual activity or other types of movement or representation on the screen. In this manner language and action are determined by how users move through the space, as well as by automatically rendered effects that connote structures within it. Any participant can activate hidden texts, and when two avatars approach each other they can begin to read each other's text. Video, voice, text and sound may be attached to objects, and objects can be made out of letters, permitting a practically unlimited range of poetic possibilities, many of which Sondheim has explored.

I have seen enactments of Sondheim and Baldwin's *Second Life* performances on multiple occasions, and each time the presentation – described by Baldwin as 'part drama, part ritual' – improves.[152] At E-Poetry 2009, with Sondheim participating remotely, via *Second Life* and a real time Skype conversation, Sondheim and Baldwin staged a chaotic intervention using their voices, mainly dialogue about their movements and locations, and avatars.[153] Both artists created spectacular vivid, elaborate and fragmented virtual spaces, characters and detritus (such as body parts), and projected these onto two different screens. Beyond the limned scenery and 'live action' that occurred, dense text passages accompanied their projections.

Presumably so as not to interfere with other actions and aspects of the scenario, text in *Second Life* mainly appears in small fonts and therefore operates in a textural as well as literary dimension. However, unlike 'Tao' and 'Dawn', text in *Second Life* is a 'noisy' experience. These out of the ordinary and somewhat difficult to read alphabetic layers add to the already ornate visual textures, including the avatar (whose head twists in a humanly impossible position), as seen in Diag. 3.16.

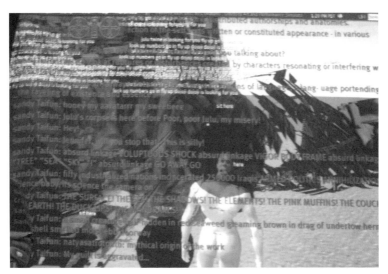

Diag. 3.16. Alan Sondheim and Sandy Baldwin, *Second Life* Performance, E-Poetry 2009.

Someone who reads the words will nonetheless glean meaning. In this segment, smaller (mostly unreadable) layers of white text that include directions, e.g. 'sit here', accompany a litany of comments in red font. The form of the segment suggests an instant message chat but without the typical two participants, only someone named 'sandy Taifun'. A sampling of successive lines reveals Taifun's informalities of speech, references to ongoing narratives, and wandering attention:

> sandy Taifun: Honey my aavatarrr my sweetieee
> sandy Taifun: Julu's corpse is here before Poor, poor Julu, my
> misery!
> sandy Taifun: Hey!
> sandy Taifun: Jennifer, will you stop that? This is silly!
> sandy Taifun: absurd linkage VOLUPTUOUS SHOCK absurd
> linkage VIGOR BODY FRAME absurd linkage 'TREE' 'SEA'
> 'SKY' 'I' absurd linkage GO AWAY GO
> sandy Taifun: fifty industrialized nations incinerated 250000
> Iraquis ARMED POLITICS ANNIHILIZATION science baby, its
> science the camera on.[154]

Taifun's non-linear statements clearly communicate flirtation and affection, salutation, commandment, distress and frustration. These parallel frenzied exhortations, which were not voiced on stage by either author, present an absurdist rejection of reality. On the whole, the audience is invited to consider multiple connotations. Sondheim wrote some of these texts; others were written by Baldwin, who describes his textual components as 'pseudo-generated' and 'flarf style'.[155] Each of the objects appearing in the scenes has their own text association. During one segment of the performance, Sondheim's 'body', 'a vast plasma', locked on to Baldwin's, the movements of which activated the stored text; as described above, actions by the characters lead to visual and verbal events.[156]

In other examples, such as Diags. 3.17 and 3.18 (below), as in many of the Sondheim/Baldwin collaborations, text and wild graphical imagery are projected. In Diag. 3.17, the texts ('G all right I live here, so what', 'come through come don't touch come in') are narrative devices adrift amid a fanciful psychedelic setting that also serve as invitations or instructions.[157]

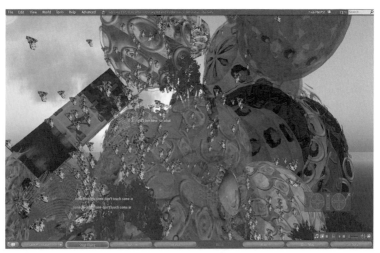

Diag. 3.17. Alan Sondheim. Detail from 'Self Induction' (2008).

Text in Diag. 3.18 is more of a soliloquy by Sondheim ('Alan Dojoji'), who types words as his avatar moves through *Second Life*, seemingly inundated and disoriented by the experience:[158]

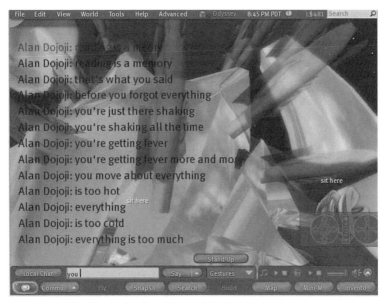

Diag. 3.18. Alan Sondheim. Detail from 'Shudder' (2009)

Apparently, the conglomeration of elements confuses and overwhelms even its creators, e.g. 'everything/is too hot/everything/is too cold/everything is too much'.[159]

During a performance, characters created by Sondheim and Baldwin dance with and against the multimedia components as a pre-recorded soundtrack plays. In addition to the spectacle engineered, the live spoken dialogue becomes a compelling part of the work in performance. At E-Poetry 2009, words spoken were part explanation and part negotiation of choreography, for example, 'can you see the other avatar?' and 'where the hell are you?'; such exchanges and questions are not only relevant for the artists but for those in attendance.[160] This aspect of the work recalls Sondheim's early video work '4320' (see *Prehistoric Digital Poetry*), in which the dialogue documented during interaction delivers a primary literary component. Baldwin's rhetorical strategy also invokes events that happen in the room, thereby bringing real space into the work so that the audience has a sense of being both a palpable witness and part of the gestalt of the event in progress. His collaborations with Sondheim, Baldwin muses, are 'dialogic, and theatrical – written as a play, with a structure of characters talking back and forth'.[161] Improvised interactions as well as texts and locations form, according to Baldwin, 'a psychodrama about power, gender relations and repression, all in a virtual environment. It's a slightly hysterical play, and there's politics because different players have different roles.'[162] A rough conceptual groundwork for these exchanges obviously exists, but they are exciting because anything – including unexpected participation by avatars logged into *Second Life* – can happen within them. Further dimensions of their project, explains Baldwin, include '[s]piritual concepts out of Tibetan Buddhism: opening ghosts and shaman inhabitations'; to this end they explore how avatars can inhabit others rather than adhere to a strictly superficial 'representational model'.[163] Through a spiritual, two-way *active* flow, the operator/user may be inhabited by the avatar to some degree. In this manner, Sondheim and Baldwin act as editors of, and collaborators in, experience; neither the artists nor their audience can ever predict forthcoming actions or circumstances.

What happens within *Second Life* are anomalous readings of virtual dimensions, the activation of avatars, and the generation of movements of objects, which Sondheim refers to as, 'the 60s gone amok'.[164] From my perspective, the work is digital poetry due to its inscriptions of text animations and mixtures of different media, especially since text plays a role and multiple digital processes orchestrate everything. While users of *Second Life* may seek a virtual interaction and not literary experiences, any image, created from the work displays literary qualities; moreover, each encounter with Sondheim's avatars or landscapes has a literary dimension. A radicalization of Charles Olson's 'composition by field' is created, in which the field is vast, malleable, and includes in the moment other people who are attached to avatars. A singular typology for this variation of digital writing cannot be determined. Many different media and attributes are at once present. A user/creator can use voice, trigger text on the screen, embed video into objects and actuate music or voice when you get near certain objects: the mediated possibilities are near endless. Objects can move and be programmed to do anything (such as respond to an avatar); a good programmer can create an object that can actually take over another user's avatar. Performance possibilities abound: you can chat on screen, audiences can come into the space and see a location from any angle, and international audiences can interact.[165]

Objects can be bought, sold, displayed, and lessons may be taught through *Second Life*'s mechanisms, but where it holds most promise for artistic experience is in its engineering of surprise and strangeness. Most users proceed within *Second Life* to take part in typical activities, such as meetings or sexual encounters. Users such as Sondheim, however, enter *Second Life* with an interest in the anomalies of that world, or what one is capable of doing within its multidimensionality that is impossible in the real world, such as flying or setting up scenarios in the air.

Sondheim's *poesis* in *Second Life* recalls but is largely dissimilar to texts previously encountered in MOOs and other virtual buildings, such as Laurie Anderson's *Puppet Motel*.[166] Capabilities of the network combine a multitude of rich media possibilities, enabling a potentially more expansive and versatile, dynamic/spectacular theatre for electronic literature. While textual delivery mechanisms in *Second Life* are at present awkward and rudimentary, Sondheim's work in this regard is likely to be indicative of virtual poetry experiences we can expect to encounter in the future. In 2008, Sondheim was awarded a residency on *Second Life*; during this period he created masterful interactive 3D works that were temporary and have since been removed. This aspect of his work in *Second Life*, its ephemeral existence, is notable yet congruent with the WWW in general. At this point, someone interested in viewing Sondheim's work must rely on screen-captured documentation, either videographic or static images.[167] *Second Life*, as such, is an unstable platform, controlled by a corporation, which will only exist as long as its purveyor, Linden Corporation, supports it.

Like Rosenberg, Sondheim invents new processes of verbal articulation in this work, which contain methodological and spontaneous elements. Simultaneously, he expands the scope of the literary to a corporeal, dramatic level. 'Tao' and 'Dawn' are lushly produced multimedia poems that contain conventional writing presented with cinematic presentation; they are more or less viewed like a video or short movie and will appeal to tastes of certain audiences. Sondheim's experiments in *Second Life,* however, are profoundly original. Creating virtual worlds for readers to inhabit was pioneered prior to the twenty-first century, but few sophisticated examples exist wherein avatars engage and navigate 3D constructs, interacting with virtual objects to trigger text and media elements. Previously, participants in such texts were on the outside looking in, activating materials. Using the mouse or other input device to control the activity of a poem within such ornate settings represents a profound shift in approach to the type of manipulation of elements occurring within the literary interface.

4

Case Studies 2: digital poetry early in the 21st century

Serge Bouchardon

Touch (with Kevin Carpentier, Stéphanie Spenlé) (2009)[1]
Loss of Grasp (with Vincent Volckaert) (2010)[2]

Serge Bouchardon earned his doctorate in Information and Computer Sciences in 2005 and is currently a professor of interactive and multimedia writing at the University of Technology of Compiègne (France). In addition to authoring the book Littérature numérique: le récit interactif *(Digital Literature: the interactive narrative,* 2009) and editing scholarly editions, he has produced educational software and several artworks.

Bouchardon's *Touch*, made with Flash, has a relatively straight-forward interface that features a hand's outline and some of its texture melded with a black background. Bouchardon describes this interface as '[s]ix scenes on the paradox of screen Touching'.[3] Using the mouse, viewers will find at each fingertip certain themed links, i.e. Move, Caress, Hit, Spread, Blow.[4] Mousing over a portion of the screen outside the hand reveals another themed link, Brush.

If we explore the content of *Touch* from left to right, the thumb ('Move') instructs viewers to '[t]ouch the words, replace them, move them.'[5] Poetry here consists of a single line that appears in the centre of the screen's stage with two slots that permute. The first slot begins with, 'Do you touch me when I touch you.'[6] Depending on which of the two active slots ('you touch me' or 'I touch you', the viewers activate, the line's content shifts. Meanings inscribed in the phrases variably confront viewers as they move through the various permutations enabled by the construction.[7] A second phase of 'Move' enables users to drag on the two active phrases to create a new line that reads, 'Do you move me when you move me'; this line may be further altered to read, 'Do you move me when I move you'; other interchangeable combinations of 'I' and 'you' are possible as well.[8]

As viewers manipulate phrases and develop a relationship with the language in *Touch*, they do so without having interference from multiple lines of text on the screen. This effect gives the work a focal point that almost demands consideration of the provocative – if slightly ambiguous – content. If someone says they are moved by something, it typically means they find it emotionally compelling. However, when someone moves something, they could simply be displacing an object by touching it. The words *touch, attract, caress* and *love* play profound poetic roles: as the viewer moves these words, different meanings of 'move' (and other words) combine, intertwine and mix together. For example, 'Do you touch me when I caress you' connotes both the inference of simple contact with the implication of a more sensual exchange.[9] These moveable phrases can be positioned anywhere on the screen but only settle at horizontal centre once two phrases co-activate. *Touch* introduces further variations of the phrase by featuring new but related themes, such as:

Do you love me when you love me
Do I love you when you love me
Do you attract me when you love me
Do I attract you when you love me
Do I caress myself when I move you
Do you love me when I caress you.[10]

Verbal constructions depend on the user's actions; they don't occur according to a prescribed order. This variable albeit limited permutation offers a range of stimulating, sensual sentiments for users to consider. Emotional congress and conflict unite through the author's clever linguistic selections and orchestrated juxtapositions. All the while, a sense of non-touch (beyond the computer mouse) also persists for the viewer.

By following the 'Caress' link of *Touch,* viewers '[b]uild a shape by caressing and following the sound'.[11] Portions of the stage feature hotspots; if you click on these designated areas, two things occur: a grainy image gradually begins to appear (although it remains indefinable until the piece concludes) and sound effects suggest the crackling of a wood fire and the sensual moans of a woman. No written text appears at all – the verbal component is the provocation represented by the woman's voice. After a certain number of 'caresses', the unmistakable outline of a female form appears, lying prone on her stomach. This section of *Touch,* in which verbal attributes are minimal, appeals to users in part because of the initial uncertainty it presents. As users participate in the construction of the picture, they have no indication of where to 'caress'; they do not know what type of image will appear and must proceed without a sense of scale or level of involvement. Each of these attributes entices users to participate in the work and delivers surprising results. With completed rendering of the woman's body, 'Caress' concludes and, fluidly, the next scene loads.

When viewers follow the 'Hit' link, they receive the order to '[h]it the fly to be able to read the text'.[12] Paragraphs extracted from Aristotle's 'The History of Animals' appear at the start.[13] This text poignantly recalls the universal nature of the sense of touch, which is experienced among humans and animals alike. In this manner, Aristotle's text operates as a potent point of departure in the work, both embedding a type of educational component and fortifying the work's overall theme(s). Among other facts, viewers learn that certain animals can live after being decapitated. Aristotle's text also becomes a visual and sound poem when joined by a calligrammatic fly, made of the letters *F, L,* and an inverted *Y,* and a soundtrack of a buzzing insect.[14] However, Aristotle's text is covered when the letter *Z* in various sizes and colours, i.e., shades of red, black and maroon, affixes to and traces the fly's flightpath and randomly appears in patterns.

When a user tries to 'hit' this fly – which is challenging, but not impossible – the sound of breaking glass and an icon representing a broken window appear at the location of the user's 'aggressive' click.[15] 'Cracking' the 'glass' makes the background text unreadable: letters and icons now cover the selection from 'The History of Animals'. The fly, when struck, turns red and 'dies'; at this juncture users may then read the paragraph and have conflicted feelings about striking the fly. In this manner Bouchardon has engineered an effective way of transmitting a conflicting experience: namely, being instructed to do something cruel and then having to face its cruelty.

If perchance the fly survives unscathed, it eventually becomes silent and the animation continues endlessly.[16] Text here, while pertinent to the engagement is, again, not exclusively the focal point. A user's actions within the decorative, obstructed construction enable them to read a text not written by the author, but one that clearly speaks to the overall premise of *Touch*: that active, sensorial principles beyond the historic and simplex point-and-click model may competently propel a digital poem. The writing may be – as it is in many digital poems – unoriginal, but it nonetheless contains a fresh artistic premise; the author has capably identified a text whose appropriation is germane to the message he aspires to transmit.

By following the 'Spread' link, users move the mouse to create a 'musical painting'.[17] In this section of the work language is absent; randomly chosen soundtracks and colour palettes load. As the audio plays, the volume/intensity of which depends on mouse placement, patches of colours, and patches of colours within these colour fields, emerge at locations on the stage where users have positioned the mouse. Essentially, users create visual accompaniment to sounds. Output here extends the viewer's response to, or participation in, sensorial qualities of the work. This scene, writes Bouchardon, 'allows the reader to *touch* the music'.[18]

In certain regards Bouchardon has achieved his aim. Users mark responses to music that the devices supply. But is it fair to call this touching the music? Objectively, the device enables users to limn a colourful response to sounds. Users have the potential to illus-trate what they hear. As in 'Hit', a user's experiences completely differ if they wear headphones – doing so exaggerates modulations of sounds, i.e. sounds decrease when the cursor is placed at the bottom of the stage; cursor shifts to the left and right balance the sound. As in *Loss of Grasp* (below), the cursor disappears during this stage. Taken together, these attributes add both sensuality and mystery to the interaction. The absence of language reflects electronic literature's variability in formation, where expressivity emphasizes tactility over syntactical reasoning. In this multi-stage narrative, the absence of words also serves as a type of interlude between verbal works, thereby giving users time to consider what they have experienced as well as to cleanse their verbal registry and conceptually prepare themselves for what comes next.

The 'Blow' section of *Touch* does not literally offer a sense of touch, but rather it presents the 'touch' of voice. By speaking or blowing into a microphone, users bring texts – randomly selected from an internal database – to the screen.[19] When functioning properly, the piece begins with snowflakes and letters moving vertically across the screen; these elements accumulate in a line at the centre. In most cases, the line of text becomes at least partially obscured by the snowflakes, though moving the mouse across the snow clears its accumulation. I have identified at least a dozen different phrases that appear in this section, which, in sum, create their own type of poem. This can be seen in the following transcription, which reflects types of intimacy seen previously in *Touch*:

What is left of my touching?
Touching with one's finger what the other person feels.
What is touching me when I touch you?
Does touching imply contact or can it be achieved from a
    distance?
I can only touch a surface.
This is my body.
Touching and playing the innocent.
Touching is private. I'm the only one touching when I touch.[20]

'Blow' offers the viewer a very attractive type of interaction: a serene scene reminiscent of blowing snow off a window to see something inside. Each line scatters once the microphone detects a sound, e.g.

blowing, speaking or typing. In this manner, the user's breath – or anything he or she physically does to disrupt silence – influences the work. Hardware and software configurations of the past would not permit such materiality on the screen; Bouchardon is among the first artists to access these new capabilities for digital writing. He uses them advantageously to address matters both of corporal and virtual tactility; seemingly he questioning the depths allowed by physical touching and revalues the level of touch allowed by the digital stroke.

The most provocative and exciting aspect of 'Blow' is how the human voice controls the poem's disappearance. The user cannot influence its content otherwise. In *Prehistoric Digital Poetry,* I briefly discussed possibilities for using vocal-response technology in digital poetry; it's obvious that Bouchardon's piece has engaged this technology.[21] For ages, poets have responded to poetry with poetry and critical commentary, but now, if we are not already capable of doing so, we will soon be able to use our voices to literally talk back to and co-compose poetry we encounter.

Let us turn our attention now to the final link embedded into *Touch,* the spatially removed-from-the-hand section 'Brush'. Interacting with this section requires a webcam; through this technology, users may, as Bouchardon states, '[t]ouch with your eyes'.[22] A greyscale, almost ghostlike image of the user's face (or whatever the webcam detects) appears on the screen amid a wavy horizontal (kinetic) morphing pattern, across which small icons (shaped as small leaves) float. Apropos of this experience, Bouchardon writes: 'Like Narcissus I can play with my image on a liquid surface.'[23] When the viewer moves her or his head (or hand or other objects), circles appear and expand on the screen. The user's engagement with 'Brush' – which relies on various technologies working in sync with each other – lasts for approximately one minute; as the viewing period concludes, the visual scenario dilates or zooms in on the scene being murkily captured by the camera. Again, no language emerges, but the user hears a droning electronic tone and something that sounds like dripping water. Through this exposure to their own stimuli – a reflection of their own ambiance and physicality – a viewer must confront their own thoughts or impressions to complete the poem. Conditions for this poem indicate primitivism, static and shapeliness dependent on modulating or controlling what an author provides. Due to this lack of potential variation (colour, shape, etc.), Bouchardon's work has limitations, but its modality indicates progress and perhaps lays the groundwork for future initiatives.

Exploring then controlling Bouchardon's structures occupies viewers of *Touch*. Non-verbal digital poetry requires its audience to be mindful of language on an introspective level. These poems conceptually transpire in an unspoken, wordless, interpersonal language contrived by the modelling offered by the author, if at all. Mechanisms in each section of *Touch* are discernible and made clear by the author; little ambiguity exists: we are told what to do. As users of the work, we either strategically or spontaneously choose a point from which to proceed; in *Touch* it makes little difference where one begins. An aspect to the work that further lessens its structural ambiguity is that it is self-contained: sections are clearly defined and an 'end' to the work may be surmised. The author sets up the same finite structures for everyone.

Bouchardon invokes Levinas, Derrida and other philosophers in his 'Presentation' about *Touch*. The thought of these theorists may provide important touchstones, but I think *Touch* can also be seen as a straightforward attempt to illustrate the tactile possibilities for sonic, visual and verbal engagement in a contemporary work of literature. 'This touching experience', writes Bouchardon, 'reveals a lot about the way we *touch* a multimedia content on screen, and maybe also about the way we touch people, objects in everyday life.'[24] Progression through the work is not static or inactive. Although its limits become obvious, the work's varied approach and possibility for closure make *Touch* a fine example of a small, effectively themed work that invokes important questions and exposes paradox. Bouchardon writes, 'to propose an artistic web-based creation on touching. One cannot touch directly: touching requires ... a technical mediation.'[25] These six scenes effectively exploit (if not exhaust) the range of possibilities currently at our disposal for sensing something via the Internet; they are commendable for successfully pursuing and staging an 'aesthetics of materiality' through the various implementations offered in an 'immaterial work'.[26]

Progressing through the work exposes viewers to conceivable forms of textual eros and roughness. While participating in the activation of content by tracing (if not creating) virtual manifestations, revelations occur as users click on or drag an object or as they use interactive microphones and cameras. Poetry as understood in a conventional sense, as crafted language used to propel artwork, is not the priority for *Touch*. Most of the lines that appear, while thematically appropriate, are neither complex nor verbally ornate. Yet Bouchardon manages to compose a clever illustration of intermediated poetic possibility; in other words, he combines elements to produce a themed poetic exercise about the experience of touch or tactile response within a literary object. *Touch* is revolutionary in that it offers ways for us to interact with electronic literature beyond simple mouse click and text-input interactions, thereby connecting more fully with the multiple senses of its users.

Now let us turn our attention to another digital work created by Bouchardon. *Loss of Grasp*, 'a digital creation about the notions of grasp and control' consisting of six distinct scenes featuring 'a character that is losing grasp'.[27] While offering different forms of interactivity at each stage, *Loss of Grasp* is entirely linear. During several moments of the work, user input renders visual output unique to that particular reading of the work, but the content and end results of each section appear similarly for every user.

At the outset of the piece, Bouchardon provides insights about how to proceed through it, though these should not be understood as instructions.[28] As the *Loss of Grasp* file loads, he establishes elements of mystery and surprise by including an audio track featuring electronically processed bird songs,[29] accompanied by a 'Please be patient ...' message that scrambles into alphanumeric gibberish if the user moves the mouse over it, an effect that foreshadows the kinetic textual dynamics employed throughout the work.

Once loaded, a single line of text, in white on a black background, sets a speculative, self-questioning tone for the narrative as a whole: 'My entire life, I believed I had infinite prospects before me.'[30] When users move the mouse over the text, the letters scramble, as described above, until the next lines – a series of passages all instigated via mouse-over – appear:

The whole universe belongs to me, I thought.
I have the choice.
I control my destiny.[31]

The poem's persona, in reflecting on life, presumes to have jurisdiction over existence, much as the user of an interactive digital artwork presupposes a type of control over the content encountered. As the piece progresses, however, we discover that neither the persona in the poem nor the user of *Loss of Grasp* controls the environment presented to them.

At this juncture, the first of several aesthetic shifts occurs: a link (via the line) is offered and a brightly coloured dot appears at the point of click. The persona's attitude remains the same ('I am the king of the world') as more dots appear wherever users place the mouse.[32] These dots serve not only as momentary décor but also illustrate to viewers how the mouse affects colour on the screen's stage; this gives the user a false sense of control. Thematically aligned passages emerge until the phrase 'I browsed beautiful landscapes' appears, and the visual and sonic scenario shifts again: as gentle keyboard notes play, users can use the mouse to create a kinetic painting – a 'beautiful landscape' that resembles a smooth (but active) form of static or colourful waves[33]. In this manner, users continue to control what occurs on the screen: shapes and patterns result from their actions. The next line, 'No wonder, because I had chosen them', again indicates the power users are led to believe they possess: that their individual choices lead to finite results.[34] In the lines that follow, however, a curious event occurs. The cursor, the arrow symbol indicating the location of the mouse on the stage, visually disappears (though its ghost still functions to guide the progression of text) and the narrative abruptly turns:

> But for a while, I have had doubts.
> How can I have grasp on what happens to me?
> Everything escapes me.
> Slips through my fingers.
> Objects, people.
> I feel I've lost control.
> For some time now,
> I expect but one thing.
> What comes next.[35]

Notably, these confessions conclude not with a question but with a declaration of inevitability. Given the shift in sensibility, users unavoidably confront their own role in the work: they have only superficially influenced it despite their constant participation. This initial sequence concludes when the time of day appears and a female voice instructs users to press one of three buttons for a 'meeting'. Button one is for a meeting in ten years, two is for one in three years, and three is for now. But only button three functions properly.

Loss of Grasp's second section begins as the 'announcer' proclaims: 'It's time for the meeting', and the text, 'But the meeting was misleading' appears.[36] Narrative foundations established at the beginning of the piece no longer apply; any sense of certainty is called into question. As the monologue progresses, via the same interactive text scrambling seen previously, one text asserts while descending from centre to bottom of the stage: 'I couldn't say anything coherent.'[37] Here, a descent into disillusionment propels the composition; six new lines appear, all but one paired with a playful homophone:

What do you do for a living?/What do you do fall and evening?
I feel we have a lot in common./Feuds we have a lot in common.
You have gorgeous eyes./You have college size.[38]

Meanwhile the work's animated apparatus – its spasticity of text – remains similar. Discrepancies between the projected text and the lines that emerge as a result of mouse-over help maintain the user's attention. The transformations of text offer surprise and enliven the narrative; the persona's dramatic distrust of anything becomes palpable. Bouchardon successfully imitates the mental state of someone whose sensitivity to language becomes a curse. The final line of this section, 'I couldn't say anything coherent', dissimilarly (i.e. non-phonically), morphs into 'I was distraught', thereby initiating yet another visual application on the stage.[39] When the line changes (by mouse-over) to 'I had to ask questions to reveal her', new visual effects materialize.[40] Either by clicking on a large (symbolic) question mark that appears or by mousing over the blank space on the right, users activate interrogative lines, e.g. 'Who are you?', 'Where are you going?'. Mouse-overs visually smear lines that have already appeared, and texts activated by both devices unveil the image of a blonde woman. Thus the 'meeting', we finally learn, involves an encounter with a woman, and the story explains, 'this stranger became my wife.'[41] The persona's search for security leads to marriage.

Nonetheless, disillusionment persists. Central to *Loss of Grasp* is the idea that human coupling does not guarantee security. Marital struggles arise in the third section. After twenty years, the author cannot determine whether a note is a '[l]ove poem or breakup note?'.[42] As this line sinks to the bottom of the stage strange sounds begin, including distorted operatic signing, and a new set of lines (eighteen in all) appear. These lines represent the contents of the aforementioned note. Bouchardon constructs these new phrases as a 3D object; mouse movement vividly alters sonic qualities, shape (proportion), location and size of the texts. These distortions recall the propensity towards permutation and variability in early digital poems: they explore how shuffling the order of lines leads to poetic result. Specifically, this segment of *Loss of Grasp* renders phrases so as to reify the above stated dilemma as to whether or not the lines comprise a love poem or a break-up note. Thus in one instance, the lines read:

I know it's a shock for you
   <<All I feel for you is love>>
    Is a lie, and
      'In a couple there is always one who suffers and one who is bored'
       I want all our friends to know that
       I don't want to stay with you
        From the first day; I have wondered how you can believe that
        I love you
         My love
          Has disappeared
          Indifference
           Is more vivid than ever
            The chasm of our encounter
             Has dissolved
             And the slightest misunderstanding
              Has vanquished
              Our love
               I know it's a shock for you.[43]

Ultimately these passages denounce love; the relationship concludes and 'Indifference/Is more vivid than ever.'[44] In another moment, the lines realign and, depending on their manipulation, can be read only partially and in reverse order:

Indifference

Has disappeared

My love

I love you

From the first day; I have wondered how you can believe that

I don't want to stay with you

I want all our friends to know that

'In a couple there is always on[e] who suffers and one who.[45]

Here ambiguity arises, thus confusing implications of the embedded language. Indifference disappears. Reversing the lines reverses meaning and intent, blurring the message communicated. Certainly, notable differences in sentiment exist between 'My love/Has disappeared/Indifference/Is more vivid than ever' and 'Indifference/Has disappeared/My love/I love you'. This section ends with the line 'Does she do it on purpose?'.[46] This flexibility of language relays the confusion experienced by the poem's persona; it indicates the type of uncertainty experienced throughout *Loss of Grasp*. For the user, however, such programmatic engineering also succeeds in enabling variable sensations and versatility in meaning, calling into question any singular line of thought or focus.

Section four of *Loss of Grasp*, a 'paper' written by the persona's son, is inescapably didactic: viewers are forced to listen to this part of the narrative, which begins, 'I don't have a hero.'[47] During its projection, the spoken words appear in block formation at the top of the stage; 'But I can't focus on the words' appears at the bottom.[48] Thus far, users have been inclined to think that *Loss of Grasp* involves the potential for disillusion in any circumstance, and – perhaps less obviously – how the user must commit to being in a relationship with the work. During the 'son's' soliloquy, in lines such as, 'the deed [of the hero] has to free itself from its creator and have a life of its own' and 'The authors' offspring will meet their own audience, occasionally finding on their way a few harsh and envious reviewers', it becomes evident that Bouchardon also reflects on the authorial struggle involved in hypermedia, specifically fretting about surrendering control of his work.[49] Once this soliloquy concludes, a single white dot appears on the stage in the field between texts, seemingly an indication of how to proceed, although deceptively so. When users click on hotspots on the block of text above it, the words' letters disperse, leaving an isolated (by empty space) new line that is not part of the original passage; revealed text does not initially appear. Subtextual messages, inscribed by the author, include lines such as 'You are not a model for me', 'We have nothing in common', and 'I want to make my own way'; these lines momentarily interrupt the otherwise cyclical presence of the original text.[50] A struggle for a representative voice and ontology ensues, which perpetuates the work's ongoing dilemma between control versus being controlled. Letters falling from the block of text scatter across the blank space on the stage, sometimes forming (or nearly forming) into words. But the section never reaches a finite resolution until the user discovers a mouse-over link associated with the last line that contains an exit.

Section five of *Loss of Grasp* returns to the line-scrambling interactivity seen previously; this begins with these lines:

Am I so little here?
So easily deformed?
My own image seems to escape me.[51]

These passages take on new meaning because users who have functioning webcams on their computer suddenly find their own live image on the stage above the line 'It fails me', while simultaneously a pulsating soundtrack plays. The image appears vividly and is interactive: moving the mouse across the surface of the stage wildly distorts the projected image of the user. By twisting and deforming this mirror image, the user becomes the central figure in the poem. Although video artists have used a video camera to inscribe an audience into a piece of art on numerous occasions,[52] this is a novel approach for digital poetry.[53] By pointing the lens of the poem at the reader, the author places the experience of reading a digital poem at the forefront. In every section of *Loss of Grasp*, users are manipulated. But the question remains whether or not the viewer – while seemingly driving the work – loses or gains control of it here. Who is manipulating and who is being manipulated? This question is not conclusively answered until the final section of the poem.

Loss of Grasp's conclusion begins by asserting, 'It's time to take control again', and '[t]o stop going around in circles'.[54] Moving the mouse on the (blank) stage, words and phrases made with smaller letters – letters that seemingly emerge from the words in the phrases – form into new words and lines on the stage, e.g. 'find some guideline' and 'shape events'. A user's input or actions seem to have some effect, but the controls – if there are any – are vague. Eventually, a text-entry box appears, but it makes no difference what the user types because the same passage appears no matter what she or he enters:

I'm doing all I can to get a grip on my life again.
I make choices.
I control my emotions.
The meaning of things.
At last, I have a grasp.[55]

The 'voice' of the poem's persona claims to have asserted control, but as users see, this control is fleeting at best. Beyond any personal connection a user may make with the work, the passage also calls into question for users the relationship they have with any digital work. At the end, as keystrokes briefly register, the user again experiences some control. This coincides with Bouchardon's forceful message: at a few moments of our lives we are in control of or may take power in a given situation; however, a sense of complete control at all times can never be achieved. Even if individuals are governed by freewill, they are guided by society and many other variables; absolute rule or power over one's own life is impossible. At best, we shape our reactions to any situation in a manner conducive to our will and attempt to surround ourselves with people that help us prosper. By incorporating common themes like love and loss – with which most viewers can identify – *Loss of Grasp* enables us to judge our own grasp on life; it helps us measure personal relationships and determine how they are and are not mirrored in the piece.

While interactive and variable, *Loss of Grasp* is linear: each section performs in a prescribed order. However, despite the fact that viewers follow a straight path, the text's discursiveness makes it impossible to predict what will happen next. This unpredictability attracts and maintains the user's attention because it inspires reflection, if not the introspection most poetry does. Being able to control the pace of the piece enables a gradual building of comprehension; the work does not change effusively. The significant degree of user involvement and effective use of sound increase the likelihood of sustaining a viewer's attention. *Loss of Grasp* is in some respects challenging or perhaps even aggravating, however, because users cannot move backwards. Once underway, users cannot revisit steps taken along the way without returning to the beginning. Bouchardon's design forces the user through a complete encounter with the work, guided by the author's prescribed path from beginning to end; a self-aborting action, e.g reloading the piece in the browser, is the only way to curtail or interrupt experiencing the piece as a whole. What's remarkable here, as in *Touch*, is the fact that within a single title, several different types of interactivity are presented: sonic, textual and visual effects are 'controlled' by mouse. This aspect of the work not only prevents its message from being mechanically delivered through tedious and redundant methods, but also requires users to think about or engage with the text in multiple ways.

I highlight this particular work precisely because of the author's attention to the sensory possibilities held by the WWW. At present, these possibilities may be minimal; nonetheless Bouchardon's work points to a broad sensory approach likely to be developed and refined in the future. Processes used in Bouchardon's digital poems vary; many include novel and unconventional interactivity. The projection of a viewer into an artwork, as in *Touch*'s 'Brush' and near the end of *Loss of Grasp*, in itself might not be unprecedented, but the fact that the viewer's physical movement (beyond the use of a mouse) immediately directs or manipulate a literary work's output is indeed a new step towards innovation in poetry.

Jim Carpenter

Erika (2002–9)[56]

Jim Carpenter is an independent software-development consultant. *Erika* was built using MySQL database technology;[57] it was launched in 2007 and removed from the WWW two years later. Of all the materials examined in this book, Carpenter's most conventionally reifies historical predecessors, even if its programming complexities surpassed most of those written previously. Because its operations and opportunities to influence the output of the program were complicated and comparatively refined, *Erika* offers an excellent example for a discussion of an automatically generated digital poem on the WWW – even if it no longer exists there.

Erika featured an ordinary interface (white text on a graded blue background), organized by links on the left, and was introduced with a quote by Robert Coover: 'What is life, after all, but a caravan of lifelike forgeries?', a message that suggests the piece copies something or that nothing is original.[58] Two sections, entitled 'Automatic Poetics' and 'Directed Poetics', enabled viewers to generate new poems automatically. 'Automatic Poetics', subtitled 'Mere unthought stuff' in the browser title bar, created abstract yet readable poems without user input: *Erika* determined the topic, form, length and word choice.[59] What can be expected of something billed as 'unthought'? Can something created by a program whose composition demanded so much time, as well as mental effort, be devoid of thought? The computer program itself does not think, but Carpenter, who worked on the project for seven years, certainly put considerable thought into it. For that reason these poems invite serious consideration.

Erika's poems do not always present a clear, focused message, even if viewers may read the work as they might any other open form poem. In the automatic poem 'A tiny peacock' (below), for instance, viewers encounter an abstract poem with recurring language and graceful motifs; other examples[60] offer similarly pronounced themes directly but often reflect the trappings of many computer poems, e.g. overt slotted structuring, irrational use of language/nonsense, repetition:

A tiny peacock

So that rest likes
   mathematics
How long would we
   be a meadow for their
   divine constellation?
We can rest
Since we are
   unknown

There is our chair,
   there is ours,
      and there the hills of white they
         put up with them

For how long might we
   be a sand
      for their odd sailor?
They and we remember thousands of
   inquests beyond us

What did our
   veins do before they dripped
      them?
Inquest, inquest, how very eternal, tiny as
   shortness, and with a
      short grace
Even though we
   wakened, a peacock were new enough
We have to guess them
Impossible, new, rapid as this noon.[61]

Quirky leaps in narrative and syntax that create abstraction
are predominant *Erika*'s poems; otherwise they are organized,
conclusive, and fairly accurate in terms of spelling and grammatical
agreement. 'A tiny peacock' is more speculative and contains some
stunning passages:

For how long might we
   be a sand
      for their odd sailor?
They and we remember thousands of
   inquests beyond us.

The word 'inquest,' by sheer repetition in the text, becomes a theme which suits the arrangement of verbal objects in the poetry, which also contains discursive stanzas that challenge sensibility:

There is our chair,
there is ours,
   and there the hills of white they
      put up with them.

One can almost hear a rhythm, even a tinge of Gertrude Stein in this passage, but the progression leads somewhere too vague ('they' and 'them') to be explicitly meaningful. Coincidental verbal connections between sections of the poem provide continuity in the litany of *Erika*'s poetry, even if the text occurs randomly. Otherwise, dissimilar voices and themes appear. The program itself cannot make the types of mindful connections that a poet might choose to make, but it does by calculated chance make illogical and attractive associations, which helps keep readers interested. The best strategy to use when approaching this type of text is to proceed as if reading a poem; a better way to contextualize an encounter with this type of device cannot be offered.

Now let us turn our attention to the 'Directed Poetics' section. In contrast to the automatic poems discussed above, 'Directed Poetics' produced radically different results. Users encountered several inter-active pull-down menus, arranged in a compact table, from which they could make choices that would affect the consistency of poems generated.[62] Entitled in the browser, 'From such rare silver', a quote by Wallace Stevens ('We do not hesitate, in poetry, to yield ourselves to the unreal') appeared above the table, which supports the premise that a poem can be created by a persona who does not exist, i.e. Erika. Before making a poem, viewers requisitely determined what they desired from the process, and then proceeded to test what the program could make.

'Directed Poetics' gave viewers the option either to use a default topic ('The death of a beautiful woman') or enter a topic of their own – although the given information states, 'there's a pretty good chance that she'll [i.e. *Erika*, or 'Etc3'] just ignore your topics altogether.'[63] One of four stanza forms could be applied: Generic Open Form, Hinge, Paratactic, and Standard.[64] The last version of the program offered thirteen 'grammars', i.e. 'syntactic templates from which Etc3 can construct meaningful utterances ... groups of TAG trees that follow some model': [Sylvia] Plath, [Rachel Blau] DuPlessis, Mimetic, nominatives, titles, Subordinate clauses, fragments, hinge clause, lyric, nominatives, [Frank] O'Hara, questions, [Gary] Snyder, and Common.[65] *Erika* contained five different types of 'word pools': topic context, topic only, topic synonyms, topic antonyms and alliteration.[66] Viewers selected a 'preferred' tense, subject and object, and also designated minimum and maximum line numbers and the desired number of stanzas.[67] Users were given two context sources to choose from: 'Complete Poems of Emily Dickinson' and 'Joseph Conrad's Heart of Darkness'. While other generators have contained adjustable parameters, the range of constraints established and enabled by Carpenter quantitatively surpassed any other experiment yet produced.

A stanza poem, made with the default settings of the program (open form stanza, Plath, topic context, present, I, her, 2, 5, Dickinson), appeared as in the following example:

Holding fear

Already I can taste fear,
   her crimson potential
It is I
   who wear her
The far eyes, opposite eyes,
   costly eyes of a
     little soul
Invisible, little, honorable as these trees
I have no hopes.[68]

Invoking obvious actions and imagery used in Dickinson's poems, e.g., 'I taste a liquor never brewed' and 'The Guest is gold and crimson', this example seems to fulfil the requests indicated to the program. A bleak, dreary and, unsurprisingly (given the title) fearful tone pervades the poem. 'Holding fear' is convincing as elementary-level poetry written in a confessional, self-expressive style.

The following example also employs default settings, with an added stanza; it also reflects a first-person narrative:

Rewarding death
I who hear my renown like
   a brave extremity
I hurry my renown, the full jealousy
   of it
Like industrious curtains
I am reddened
   by an exclaim

Rare as a privilege
I should be a
   deer.[69]

Beyond the vocabulary of Dickinson, the isolated, metaphorical but self-oriented voice clearly indicates Plath's influence. As in the previous example, it's easy to identify unusual words that can be traced back to Dickinson, such as use of the verb *redden* (see poem #342), which is, coincidentally, a prominent word in Plath's poem 'Moonrise'.[70] Observing insertions of different voices and other parameters, readers of *Erika* in some sense succeed in constructing a Gary Snyder et al. poem with the words of either Dickinson or Joseph Conrad:

Living potential
Already I can taste privacy, my scarlet
   potential
Since I rejoin her
Fun, consummate, stiff
   as these stitches
The women scream

Lives and forestalls, but
   there is no immortality
    beyond this hunter
Somewhere a frost is shorter
This is what it is like
   to be sweeping –
    so perished

To crack a gross cycle, still
   existence, an unopened heaven, poverty, a plummetless
    dew, an ample psalm
They can be a
   man
A dry business
   flowed.[71]

In this sample, the sensual, masculine style of Snyder appears, replete with sexual innuendo, e.g. 'Since I rejoin her/Fun, consummate, stiff', and natural imagery, as do Dickinsonian phrases and word use, e.g. the formation 'there is no …', #1263; use of plummet, #663.

Importantly, titles and themes in the above examples connect without much difficulty. The syntactically correct language, while logically disconnected at points, does not defy poetic linearity. Since the Modern era, readers have had to contend with discursive verse. In comparison to many earlier examples of computer poetry, *Erika* exhibits a refined coherence. Carpenter, in his complex program design, managed to find a way to minimize repetition and engineer variation. Each example shown above builds its first lines around the pronoun 'I', as requested. From that moment we find no such similar patterns; likewise the program produces a significant range of grammatical structures and punctuation use.

The output's appearance does not greatly change with different parameters and, given the gloom of the 'context sources' (Dickinson, Conrad), neither does its tone. When drawing from Conrad, bleak themes continue to appear. In the next example, with O'Hara's grammar and a 'preferred subject' of 'we', the aesthetic and thematic qualities compare to those above, with overt imagery, e.g. 'rivers', 'rot' and 'torchlight', appropriated from *Heart of Darkness*:

Generosity turned into nighttime
Creeks turned from gloom
My hands wild
    with rot
This dark may remember and return, but
    it is utterly tawny, rays,
        expressions, rivers, the
            rolling faces

What can the arms do without
    arm to dishonour?
I can stay
This arm may set and thicken, but
    it is bitterly imperceptible

Supreme as a time
Here is a mangrove, a mouth,
    an invasion, courtyards for a disciple
Pensive leaves and merry doors
They may be an
    invasion
To exchange tenebrous
    stones and immense brutes

Like a virtue
That they writhe her
In left they know
    a glance, coming
        beneath her post, contorted from hush
Into a swept branch a
droll destiny blunders
A slime of
    her flying occupies a terror to
        a young torchlight of nature.[72]

One does not particularly encounter the casual, conversational
style of O'Hara, and, curiously, the requested pronoun is ignored.
A sense of peril appears – the continuous danger that stalks the

individual and humanity reflected throughout Conrad's works – but the program forsakes other input parameters. Depending on the user's threshold for poetic abstraction, e.g. 'Supreme as a time/Here is a mangrove', 'That they writhe her/In left they know/a glance' or 'A slime of/her flying occupies a terror', a viewer may not derive meaning from the output, even if it falls within the boundaries of accepted poetic licence. Adjusting factors that fed the output brought similar results. *Erika*'s poems often ignored the 'preferred subject'; they tended to deliver results that seemed more focused on the default and consistently featured a type of vague meaning. In certain respects, the program could not be relied on to perform flawlessly.[73] However, since the same sets of words do not consistently repeat, *Erika* maintained a type of freshness historically uncommon in computer poetry. In the event that *Erika*'s language became stale for readers, vocabularies could be switched to employ a different set of words.[74]

Erika consistently performed as above. Altering aesthetic areas, such as her 'grammar' and 'pools', demonstrable differences in output emerged, although input topics, subjects and objects were not always recognized. Only if *Erika* identified a word in the suggested topic would the output reflect a viewer's suggestion. Through adjusting the technical rather than linguistic parameters, working with *Erika* became thoroughly engaging; becoming surgically involved with the code, perhaps adding vocabulary to the context source or creating additional context sources or grammars, even deeper levels of engagement could be reached.

In co-opting vocabularies from a single writer or book, Carpenter imposes poetic constraint. Incorporating stylistic traits specific to a known writer, however, adds expanse – if not intrigue – to this work. *Erika* certifies the plausibility of digital mimicry, and this was her great success. Results of this reader-program interaction invite literary criticism. *Erika*'s vocabulary and program supported formidable poetic output that pointed towards the disconsolate side of the emotional spectrum: *Erika* spent most of her time portraying a tumultuous existence, no matter what users requested. A computer poem relies on the strength of its input, dictionary and/or database – and even a stellar database does not guarantee success. *Erika* less successfully emulated the grammar of writers not known for alienated themes, like O'Hara. Yet by incorporating language and styles by writers who explored similar emotions, like Dickinson and Plath, who had entirely different expressive approaches but covered similar ground, the program succeeded in cultivating a hybrid voice.

Erika's intriguing manifestations compare favourably to the most sophisticated historical models, e.g. poetry generators built by Jean-Pierre Balpe. By varying the appearance of stanzas and minimizing the presence of repetition, permutation and generic slotted structures, *Erika* excelled. The words of one writer chosen by another writer/programmer created unique poems capable of attracting and maintaining a reader's attention.

However, a computer program challenges us to consider the plausibility of artificial (or, to use Stevens' term, 'unreal') text. One could be mechanically satisfied by *Erika* – a Frankensteinian cyborg poet – but not moved by her verses. As a persona, she imitated well, acquired functional grammar, and showed flexibility within her expressive formations. On the other hand, she did not, and could not, always give us what we asked for due to her lack of real cognition.

Researching the topic for nearly twenty years, I have encountered dozens of poetry generators.[75] Many have issued convincing poetry, but even the best of them fatigue the reader with blatant slotted structures and repetition. Computer poems result from many actions, enabled by responsive programming fundamental to composition. Although *Erika* didn't permit users to add text and bridled authorial command, few programs are as nuanced and versatile. *Erika* possessed attitude as a virtual author. Sprinkled throughout the site, quotations constellated the author's perspective on the poetry industry. Samples of statements by Charles Bukowski ('poetry is still the biggest snob-racket in the Arts with little poet groups battling for power'), Allen Ginsberg ('Nobody publishes a word that is not the cowardly robot ravings of a depraved mentality'), Theodore Roethke ('Delicate the syllables that release the repression') and Gertrude Stein ('How can you tell a treasure. We can tell by the reaction. And after that. And after that we are pleased'), holistically cast a critical eye on literary pretension.[76] Recalling the message transmitted in Coover's epigram, viewers could infer that *Erika* served as commentary on authenticity and contemporary poetics. These epigraphs also suggest further contexts for *Erika*, relating to the reality of mechanical craft and sensitivity in poetry and poetry's social identity. *Erika*'s objectives and successes can be gauged from these touchstones. Personified as a virtual author, she generated plausible poetry, proving herself to be worthy in this regard by giving readers many variables to adjust as they construct a computer poem, and then varying the output widely.

Angela Ferraiolo

Map of a Future War (2008)[77]
The End of Capitalism (2009)[78]

According to a biography appearing on the WWW, Ferraiolo, whose professional background includes working as a teacher but also working in the Gaming, Theatre and Broadcasting industries, 'is interested in the re-imagining of existing literary forms that can, through the digital arts, be exploited for their liveness and that perhaps were never fully realized while they were written for print. She also envisions a move towards a kind of code process that can be 'literary'.[79]

Ferraiolo classifies *Map of a Future War,* published in the online journal *The New River,* as a 'Spatial narrative./Repeated access of a character set as data.'[80] In *Map of a Future War,* she engineers, using Flash and JavaScript, a visually demanding poem that reflects the refined attributes of WWW-based literary hypermedia.[81] Ferraiolo diverges from forms and presentations of interactivity offered by Bouchardon, despite the fact that mouse-over figures prominently in *Map of a Future War.* Her compositions feature apparent and cloaked characteristics, and much less linearity than Bouchardon's work. These facts bring added challenges for viewers, who, in addition to poetic language, face beguilement through the interface.

Ferraiolo's description of *Map of a Future War* as 'a character set as data' is telling and useful for understanding the overall conception of the work. The 'set' from which she assembles the work contains letters (words), symbols (icons, etc.) and no sounds. In essence, viewers ultimately discover Ferraiolo has prepared a device that creates and assembles a series of collages containing interchangeable, discernible, separate attributes and texts. These collages fuse fluidly through hyperlinked coloured hexagonal symbols. Although an overt design logic exists, it is not easily detectable.

At the outset of *Map of a Future War*, Ferraiolo provides no instructions. A full review of the interface reveals key elements. The work consists of a few words (e.g. 'product' and 'path') and phrases (e.g. 'lies=lies', 'hourly wage', 'we accept Euros' and the title) collaged with a numeric chart and schematic-like symbols atop a grey background.[82] Viewers also find three hotspots. Clicking on the title, which appears above the phrase 'money means everything', initiates the main body of *Map of a Future War*; two small icons at upper right can be used to enlarge the window to a full screen and/or to call forth an alternative version of the opening interface, one that features the outline of a woman's face (in green) along with the word 'Route' and iconography similar to its counterpart. From these opening screens viewers can begin to deduce some of the business-related themes of the work.

Like many examples of digital poetry, *Map of a Future War* starts at the same place each time, and then branches out into sequences unique to each viewer. At this opening, viewers become familiar with the dynamics and tenor of the poem. First, in orange and of largest scale, is the pronouncement: 'first drink of the day a killer'.[83] Beneath it, Ferraiolo presents four layers of readable yet 'faded' text in different colours, which come to the foreground via mouse-over, while the other texts remain visible in the background; these lines overlap and contain decreasing scales of font:

mindlessness chatter been at it all
day the flat of his hand sweeping
down the counter none of them
able to explain or rather think
therefore continuing all this balled
up spit out as news Christ so much
fuss over money (in blue);
'but nothing came out right/not that last time I saw you' (green); and,
a
correction
he
called it (red).[84]

In these initial passages emerges a hierarchy. Although not always present throughout *Map of a Future War*, this hierarchy effectively introduces several themes inscribed in the work.[85] At the top is the introduction or pronouncement of escapism through unhealthy activity, i.e., drinking – a trope historically advanced by numerous writers.[86] Apparent in the second and longest passage is the prevalent anxious atmosphere of the workplace, transmitting the ironic insecurity associated with certain types of jobs. Subsequently the text establishes the instability of interpersonal – or perhaps any type of – communication, as well as the stoicism of the language of authority. Unveiled passages, individually and as a whole, formatted with line breaks, inscribe a literary component that provides overt written connections with poetry as a historical form. Visual and verbal fragments, as well as the kinetic movement occurring between passages, support the primary verbal expressions; each part is integral to this digital poem. On the opening page, features include the following: hand-drawn diagrams connected to the words 'Path' and 'Route'; a line drawing seemingly containing a face on the body of a fish; a MasterCard logo; dotted lines connected to words (or fragments) such as 'one', 'be', and 'was', and the words (or fragments) '(Deal Breakers)', 'IPO' and 'Dow'.[87] These elements introduce a degree of visual noise, but crucially serve to solidify the foundations of Ferraiolo's work.

Although *Map of a Future War* imposes no temporal constraints or elements, since viewers guide the pace of movement and moving through the poem presents little difficulty, it is not without compli-cation. After a brief scan of the opening screen, it becomes evident that hexagons are hotlinks and that other text fragments, drawings, and symbols are temporarily fixed compositional dimensions. At times these elements depict not only corporate iconography and logos for companies that have control of the world, but images associated with nuclear weaponry, petroleum power, global warming and animal extinction – all players in the game of human hatred, hostile behaviour and war; they also illustrate the hazy cloud of infor-mation people are exposed to throughout society.[88] Beyond these basic constructions, Ferraiolo plays some deceptive tricks. While hexagonally shaped hotspots are obvious, other symbols, such as arrows and lines connected with the words 'route' and 'path', appear to contain some sort of guidance but do not function as active links; a system for interpreting these notations is not established, as in Rosenberg's work. Diagrams included are often incomplete or inconclusive; these suggestive paths objectively lead to furtive, ambiguous ends – deciphering them is impossible in a conventional sense. They possibly mean to be ambiguous representations, intimated directions that don't lead to finite destinations. In addition, viewers cannot retrace their steps through the work. While numerous forward paths are possible, Ferraiolo disables backwards movement through *Map of a Future War*; as a result, it is as if the textual record or organism has no memory and cannot be repeated.

Map of a Future War contains varying degrees of visual complexity; while not a game, it contains playful, colourful characteristics and demands viewer interaction. The frequent overlap of contrasting colours, text and icons make the work challenging to absorb seamlessly. Words and symbols often (though not always) obscure each other, even when brought to the foreground via mouse-over, although viewers are always given enough information to receive the poem's content adequately. The author uses the framework of the stage cleverly, often affixing words at the margins so they appear as topically related fragments, e.g. 'route' becomes 'rout' and 'down' becomes 'dow'. Overall effects inscribed in Ferraiolo's layout effectively engage viewers with variety; though this layout is cluttered in places, it remains aesthetically effective as a multi-layered collage. Each aspect of non-obstructive noise contributes to the work's overall message.

In terms of links, viewers always have a choice of multiple paths. Each screen appearing in *Map of a Future War* contains eight hotspot links to fixed textual endpoints: clicking on the same sequence of hexagons will lead viewers through the same passages of poetry. The activated navigational apparatus is impressively unique: when viewers click on a link, the entire stage slides up, down or diagonally as a complete unit until the new set of (similar) elements settles into place. All the map's materials appear as if magically laid out on a large shape-shifting table. Readers glide and pore over Ferraiolo's virtual grid.[89]

Iconic symbols and text fragments (described above) – fixed in place and formation – may appear several times during a single encounter with the work. By passing through Ferraiolo's map and seeing previously viewed arrangements (templates) of information, viewers recognize that while hexagon links are consistent in number and placement, active texts vary in number: as many as eight appear and as few as three; viewers also notice that these change over the course of a reading, i.e. new passages are inscribed. All of *Map of a Future War*'s links are pre-plotted, but many different paths through the work – each containing unique passages – exist. The quantity of screens and lines comprising the work depends on which routes viewers follow; this gives the work a sense of indeterminacy.

Due to thematic consistency, arbitrary choices are as effective and efficient as any other approach to accessing content. Themes established at the outset reliably appear in every sequence of *Map of a Future War*. In the example presented below (Diag. 4.1), amid the banking, media (CNN), communications and other iconography, viewers find passages:

> or waving the polluting
> smoke
> away until the next
> of us
> lights up
> too then another and
> another and.[90]

This passage combines with:

> standing in the centre of
> the floor his arms
> outstretched his eyes
> looking up continually
> reading the numbers that
> hang above our heads
> like a lamentation like a
> message we might have
> found scrawled across
> the pink bricks of the
> most beautiful building in
> the capital just as the city

was about to encounter
another catastrophic
event reading watching
he repeated his offer.[91]

A final few lines are also part of the piece: 'come up to the office he said/his voice flat unemotional pure/information'.[92] While a sense of communication's failure does not explicitly emerge, as above, the gestalt of the work – cultivation of an unhealthy atmosphere, monotony, and stoicism in the language of authority – persists, juxtaposed with the inclusion of a nostalgic passage:

I will miss the photographs
posters toys that filled the
cubicles the simple
morning exchange of
news gossip opinion
jokes in the elevators.[93]

Ferraiolo constantly reiterates these themes throughout *Map of a Future War*, establishing intention and enabling viewers to receive an unhindered message.

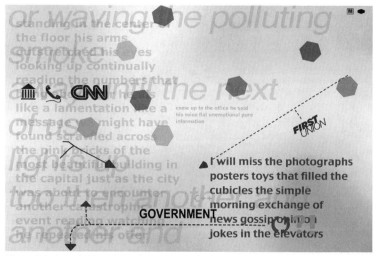

Diag. 4.1. Angela Ferraiolo. Screenshot from *Map of a Future War*.

Reading the passages more closely, it becomes evident the poem has multiple purposes. In addition to questioning the integrity or envisioning the failure of an economic system or company, scenarios of commercial breakdown also subtly acknowledge collapse of a relationship in the persona's personal life, which sometimes corresponds to or overlaps with the general theme of experiencing difficulty in all types of communication in business environments.

Map of a Future War thus does not limit itself to existing as an artwork about the injustices of business or the deception and complexities of numbers, the miasma of trade. Ferraiolo also acknowledges human failings and grief outside the realm of commerce, thereby suggesting that these collapses may be related. Nonetheless, the majority of poetic passages unmistakably focus on crisis in the workplace. This set of associated lines, like many segments of the poem, masks or excludes altogether tribulation within an interpersonal relationship:

> woman at her
> desk muttering
> under her breath
> bastards the
> bastards when
> I go to speak to
> her she starts
> to whistle
> ...
> nobody
> to feel sorry
> for since
> you can't
> reasonably
> feel sorry
> for them
> ...
> talking like
> one of
> those tent
> city

preachers
deluded
self-
righteous.[94]

These excerpts project the anger, sorrow and delusion experienced during crises in and out of the workplace, even if the identifiable subject (a woman at a desk) suggests a professional environment.

Map of a Future War ends when viewers confront the 'alternative' opening interface and can then, from there, begin again. In general, a viewer sees ten to fifteen screens before reaching this endpoint, after being exposed to approximately forty or fifty brief stanzas of text and accompanying décor.[95]

A line appearing in one of the screen templates, 'WHAT IS HIDDEN/FROM VIEW', coincidentally joined with 'The Unavoidable', 'Too Late/Too Late/Too Late' and VISA credit card icon, provides a key to the work. Beneath professional and economic systems threatening to collapse, personal experiences full of complexity and difficulty also exist. Rather than putting the latter realities in the foreground, Ferraiolo keeps them behind the scenes. This particular example shows how lines between the two become blurred. Given all the suspicion of numbers and deceptions inherent in business portrayed in the poem, these fragments could be interpreted as a critique of unsavoury elements that drive economics, but can also be seen as a comment on the undercurrent of troubled personal relationships. Ferraiolo's aesthetic and design choices in the piece also tie in with this encapsulation of the work's identity as something being hidden from view: none of the screens prevents viewers from reading or seeing anything, but overlapping text and icons make reading the material challenging.

Due to navigational intrigue – such as the inability to proceed backwards, or being enticed to investigate logical patterns in link colours while reading the work – *Map of a Future War* presents a strong example of how mediated mystique can be cultivated. Flexibility and variability, combined with the huge number of possible paths through the work, its textual variables and constraints prepared by the author, prevent all but obsessive viewers from absorbing all the materials. Long before someone looking at the work could possibly do so, Ferraiolo's design forces them back to the beginning. Some content is likely never to be seen, but this does not matter because of the unified focus in the work, where every viewing captures its thematic threads. Anyone who spends time engaging with *Map of a Future War* leaves with a similar message, no matter the duration of their investigation. A political implication regarding the role money and class – further explored and sustained in Ferraiolo's *The End of Capitalism* – is an overt concern. Banking, commerce, credit, corporate ontology and the uncontrollability of money have come to dominate human existence, a fact not lost on Ferraiolo, whose persona in the poem questions these systems and portrays the damage and internal suffering caused by them. Beyond the mundane (or tragic) experience of working in an office or cubicle, personal turmoil arises.

At one juncture in *Map of a Future War*, the line 'But calm in concentration in strategy the screens carrying the news' appears.[96] Ferraiolo develops a strategy wherein her screens do carry the news. Language and the presentation of language in *Map of a Future War* represents a type of *poesia vérité*, or true poetry (poetry truth), even if it is fabricated by an author who imagines dialogues and thoughts that could occur within an office place scenario. Ferraiolo combines realistic, plausible occurrences with stylized poetic devices of linebreaks, colouring, layers, symbols and programming; she renders collages and uses the WWW's capabilities to invoke and reflect lifelike subjects. She takes a provocative stance towards her topics and finds a dystopia. By imposing paths, she channels the work in a direction that offers few possible interpretations but gives viewers countless ways to reach conclusions. The work's title suggests events to come, and as such could be considered an attempt at prophecy. However, as the content accumulates, as the language portrays snapshot visions of economic collapse alongside corporate logos and individual failings, we realize this is a poetic dispatch on current events: this is not the world we are heading towards but have already arrived at. *Map of a Future War* was published during a global financial crisis (Autumn 2008) and reflects capitalism's influence on contemporary realities. Written passages on any thematic level, presented from an observational point of view, may or not be real, but the scenario rendered – representing struggle between, and common objectives of, employees and corporations and how both cannot win – sensitizes anguish caused by the failures of large business institutions and personal relationships.

In *The End of Capitalism*, an interactive videographic poem both responding to viewer input and progressing on its own, Ferraiolo explores new aesthetic territory. The work focuses on 'economic beliefs', and represents one of first artistic efforts to integrate – formally and directly – many if not all of the WWW's capabilities within a single piece of electronic literature.[97] She does this by using Flash to combine video (found films, cartoons and her own footage), written text passages, hyperlinks and animations – thematically united by their focus on, and considerations of, the concept of capitalism. Ferraiolo accomplishes the task through 'the coding of a movie engine that models character states, then assembles the montage'.[98] Whereas ordinary movies follow 'causally linked narratives', Ferraiolo explains, 'this kind of movie is made up of episodic sections that can be interrupted at different points and sequenced in different orders.'[99] While the process of modelling character states remains undefined, presumably this means her program analyzes or evaluates viewer input and positions and sequences written and mediated passages based on choices made by the viewer. Directions are not provided, so it is unclear what to expect from *The End of Capitalism* at the outset, other than it is interactive.[100]

The End of Capitalism's opening sequence features a montage of two black and white videos, one kinetically rendering (scrolling) newspaper stock reports and another featuring a group of people actively at work at a stocks and bonds company, while the title scrolls horizontally across the screen on different lines. Viewers are introduced to the omnipresent ambient soundtracks – most often a droning electronic pulse – that frequently accompany text and video passages and set an aural tone for the piece. Intermittently, at lower right, a small animation featuring six coloured dots appears, indicating the viewer's singular point of entry and engagement. This symbol – a button consisting of coloured circles – also represents the interactive area of the videographic segments and appears in different iterations throughout the course of *The End of Capitalism* as a navigational lever.

Once viewers become familiar with how the piece functions, the interface presents few difficulties. Ferraiolo prohibits pausing the work and moving backwards; viewers cannot conveniently return to a place, screen or animation already seen, so decisions on which of the two paths to follow, linking or automatic, have significance. Because of its indeterminate values, each viewing of the work, though united with other readings due to thematic similarities and sharing passages from a common database, is unique. Throughout the work, Ferraiolo presents film clips, sometimes juxtaposed one atop another (a smaller video layered on top of a larger video), interspersed with text passages appearing atop a background of flashing dots (similar to those used for the buttons). To its credit, *The End of Capitalism* will proceed on its own, independent of viewer input; we are not required to do anything but watch and read what appears on the screen. Simply letting the work run its indeterminate course enables different sequences of different lengths to transpire. However, once the piece commences, media content is activated via clickable buttons and screens; sometimes buttons do not appear immediately, and a certain amount of content must be viewed before a viewer can progress. If someone watches *The End of Capitalism* several times, textual events will repeat, although scenes and text passages do not reappear within the course of a single reading. The reappearance of the opening screen marks the conclusion of a viewing of the work. Ferraiolo does not impose a formula for reaching an endpoint or common culmination. Viewers might coincidentally see the same or similar content as *The End of Capitalism* draws to a close, but a fixed path to the end – a preordained concluding sequence – is not programmed.

The End of Capitalism, with its copious videographic materials, contains considerable amounts of animated visual activity; it thereby represents a culture that resists being construed as a static entity. As mentioned, even the backgrounds of text passages are kinetic, appearing as a series of animated dots, larger scale versions of the interactive narrative buttons. This characteristic seems largely decorative, but also acknowledges a steady stream of behind-the-scenes activity, and refers to the presentation of information on televisions in the past as a composite of RGB pixels. Ferraiolo presents some text passages kinetically, one line appearing at a time, an effect that results in textual and visual formations that read slightly ambiguously and dramatically, more like poetry, as in these two segments:

> hearing it on the radio
> pulling off the road
> getting out of the car
> trying to catch your breath
> not hearing the oncoming traffic
> why were they saying this
> not seeing the headlights approach swerve
> that's not what money is;

and,

> part of a fantastic machine
> never ask for help that's my principle
> the worst thing is the constantly cheerful news
> things not people and things not ideas
> it isn't everything
> it's temporary
> what is it I believe.[101]

The first passage implies tragedy and disillusion caused by capital. For someone who believes in the system, money supposedly brings good things, but so many times the destruction caused by capital dominates news stories and results in personal crisis. Ferraiolo's line-by-line unveiling of the passage adds drama, if not progressively iterated desperation, to the work. As in Sondheim's 'Dawn', kinetic presentation steepens the emotions expressed, and temporally renders a dialogue alternating between euphoria and dejection. The second passage more enigmatically acknowledges the awe and selfishness accompanying our commercial system – its common emphasis on the importance of goods – while simultaneously recognizing its weaknesses and ephemeral construction and the disenchantment it engenders.

Ferraiolo appropriates video samples from archival footage of various sorts; many clips were produced in order to celebrate the benefits of capitalism, and others reflect its inequitable results. Despite a possible imbalance between the two, in favour of the former, no mistake can be made about whose side the author is on. By including her own footage from Wall Street protests and congressional hearings on the collapse of corporations such as Lehman Brothers, Ferraiolo clearly critiques and demonstrates contradictions within the current economic system; her counterbalance is poignant and not subtle, if not didactic and heavy-handed. The scene where a corporate CEO testifies at a hearing to clarify his assets demonstrates the arrogance of big business; this information provokes the viewer to realize how those at the top of the economic food chain look down and prey on those below. To her own benefit, Ferraiolo wisely includes less blatant symbolism in several of the videos that populate her database. I am more moved by the simple, soundless footage of an impoverished family riding together in the back of a pickup truck, or the silent ruins of an abandoned building, which are more subtle and poetic than watching white men project propaganda, consumers fondling products, or a woman looking at herself in a department store mirror. Poignant moments occur as a result of Ferraiolo's selections, combined with results of the behaviour of coding she uses to propel the work. Because she avoids the use of linear, pre-plotted sequencing, viewers can never anticipate what will happen next within the narrative. The combination of peripatetic modalities, variety, discrepant contents and surprise produce a pleasantly unsettling, moving emotional effect.

Video portions of the work present consistent messages. Their content predominantly consists of capitalist and 'educational' propaganda relating to the promotion of popular national (now global) values. Focusing on the United States, *The End of Capitalism* questions given realities of contemporary cultural conditions. Ferraiolo incorporates strictly American videos, reflecting native ambition (e.g. Levittown), traditions (e.g. Thanksgiving), and then the consequences, mixing commercial iconography (propagandistic, persuasive) with jingoistic (voiced) opinion, commentary and sentiment ('everyone can be rich,' 'everyone is happy').[102] Capitalism dominates our lives, and its downsides – portrayed in one scene by footage of people protesting for more jobs, in another with the chant 'bankers go home' – can be devastating. Loss of a job or even the need for a job can be ruinous to the quality of one's life, not only causing financial turmoil, but also social agony, as people without jobs are sometimes belittled and stereotyped. Ferraiolo's intermittent weaving of videographic segments demonstrably reveals and underscores the range of social conditions brought on by the role money plays in our culture.

Expression broadens and humanizes in text-based passages. Artful language elevates the work's scope to the literary. These interventions, emanating from a persona's recollection and thought, poetically occur in the work. As digital poetry encompasses numerous forms of expression, these texts thematically support, and are supported by, the cinematic elements; Ferraiolo carefully crafts a discrepant but intentional relationship between the two types of presentation. Text passages appear automatically either when viewers click on a button or as dictated by unexplained operations inscribed in the code.[103] She devises a few different types of text passages varying in format; some are static, some conventionally animated, appearing line by line, as above. Many appear in paragraph form with words and phrases in capital letters as in this example:

and without it HOW TO LIVE the stories a song that will not fully come to mind a group of pilgrims walking hand in hand like actors A MYTHOLOGY some image lighting up behind the billboards at the bus stop while waiting in line at the bank VIVID then fading dead smoke on a mirror then APPEARING AGAIN out of the blue implying this is the way somehow we knew it all along gesturing coming through the door walking in sitting down smiling confident then instantly gone again a FABLE nobody grasps or can put into words made up or waiting to be rediscovered retold starved for attention turning to look at us with that astonished sense of TOLERANCE and PATIENCE thinking how could it be like that or thinking WHAT.[104]

In this style of passage, emboldened words telegraphically encapsulate the ornate prosaic message otherwise inscribed. Text formatting choices distil a message. Further, using a stream-of-consciousness and distillation approach, combining memory and imagination, Ferraiolo offers a vivid narrative in which the poem's persona looks for ways to make sense of, and create alternatives to, cultural conditions by making life more than ordinary (mythologizing), finding surprise and embracing incompleteness. Other long passages using the bold formatting technique question history and tell impromptu stories.

Another type of passage, without formatting, confesses confusion and regret, as in this segment:

> or maybe it had never been real never meant anything maybe
> the money had never been there lost on champagne private
> helicopters hand-carved cigar boxes the grayish petrified
> substance constituting each atriumed tower their monolithic
> windows absorbing light sound doors so huge they engulfed
> everything deliberate arrogant a condensation of arrogance
> Cayman Island estates South American divas oil paintings
> rock stars unblended scotch gracious donations to the
> symphony private schools super model girlfriends sports
> cars couture third-world servants bulletproof limousines
> home movie theatres box seats golf courses emeralds.[105]

The End of Capitalism, while not completely one-sided, on the whole responds to the evils of excess; the above passage more than implies apparent results from inequities now present in society, here railing against decadent lifestyles. Ferraiolo's critical apparatus is in the open, contemplative and ruthless, yet leaves room for discussion. Static passages without boldface, though seemingly constant, appearing the same way every time, actually contain minor shifts, depending on browser size. They also contain a Moebius quality; the order of lines seems inappropriate to the larger story they are part of, in which we live in 'the best country in the world', but lament the passing of an era when 'a dollar was a dollar'.[106]

Ferraiolo employs a range of textual design techniques. In another type of passage with bold features, she presents a more complex composition:

soon it would all be
relegated to cinema
advertising a page long
editorial NOBODY READS
ANYMORE not since the
last war its social
function currently
REVEALED and a bit too
nakedly reminder the
richest country in the
world one person after

SPITEFUL when anyone
must admit there is A
CYCLE turbulent
explosive and no one
controls it there being
LITTLE PROTECTION in
any philosophy or under
any system of business
that protects the common
individual from greed
mistakes POOR TIMING.[107]

Like many Modern-era poetry experiments, beginning with Mallarmé's
'A Throw of the Dice Will Never Abolish Chance', design of layout
here offers viewers multiple ways to read – either horizontally across
the screen or vertically down both columns, or as two separate texts,
as well as a third made by bold words.[108] Obviously the passage
was originally written vertically, with imperfect but linear syntax. A
conventional reading of the passage offers perhaps an explanation of
Ferraiolo's own turn to cinema, e.g. 'NOBODY READS/ANYMORE'.
It also reflects and recognizes troubling systemic effects of advertising and business congruent with themes presented elsewhere
in the piece, e.g. anyone/must admit there is A/CYCLE turbulent/
explosive and no one/controls it there being/LITTLE PROTECTION.[109]
Alternative approaches to absorbing the text, however, expand its
range; they invoke emotion from the first line 'soon it would all be
SPITEFUL', and in fact criticize the prospects of cinema, changes in
literacy, and human manipulation brought on by a culture absorbed
by media, e.g. 'anyone/relegated to cinema must admit there is A/
advertising a page long CYCLE turbulent/editorial NOBODY READS
explosive'.[110] Ferraiolo's mode of presentation contests contemporary
expression, challenging viewers to explore and find different sensibilities within her words, and brings further versatility of expression
to The End of Capitalism. This alternative perspective on the composition reveals mistakes perceived by the artist, possibly suggesting it
is a bad time to curtail reading. In all, The End of Capitalism contains
verbal passages poetically operating on more than a single register,
which add to the depth already attained via media complexity.

Creative sophistication in *The End of Capitalism* results from atypical coding and processes i.e. hypervideo, but also from the research conducted by Ferraiolo, who goes to great lengths to find relevant film, video and television footage to embed into the work. Beyond these tasks, she shoots video of her own and composes the written passages. These elements, integrated by and through the code, offer an intriguing experience for the viewer. What are the differences between active and passive viewing? Is watching *The End of Capitalism* in a passive state, not clicking on the buttons presented, as efficient and suitable as engaging with the work and altering the algorithmically divined path? I do not believe active users get any more, or special, content by interacting.[111] Because only brief fragments of the videos appear at times, variable quantities and qualities are enabled even under passive circumstances. To make a stylistic literary comparison, if viewers hear and see the audio and video components in *The End of Capitalism* as poetic lines, as they should, the condition of text that curtails content essentially erases details (at least partially) and therefore enables the work to expand in subsequent viewings. If this were some type of magical book, the second reading of the poem would feature additional lines on its pages. Of course, historically this would be impossible physically, but with tools currently available, materials presented may be altered improvisatorially according to computational rule.[112] Alternatively, interacting with the media provides a sense of participating in a personalized construction of the work, an attribute benefitting viewers who seek engagement and who wish to customize their artistic consumption.

The End of Capitalism shares aesthetic and to some degree mechanical similarities with Lev Manovich's *Soft Cinema*. Manovich has invented a process for making computer-driven films developed in the twenty-first century in which, he writes, 'human subjectivity and the variable choices made by custom software combine to create films that can run infinitely without ever exactly repeating the same image sequences, screen layouts and narratives.'[113] In Manovich's project, software selects video segments from a database of footage and performs different sequences and edits of the footage. Each performance differs because the software enables every dimension of a film to change. A type of certain indeterminacy found in Manovich's project corresponds to effects found in Ferraiolo's interactive videopoem, though *The End of Capitalism*'s production and interface are not as visually complicated as *Soft Cinema*'s. A mystique develops and is then withdrawn during the exchange between the viewer and artwork in *The End of Capitalism*. After a certain threshold is reached, viewers begin to build confidence in their understanding of the poem's organism; its aura – any sense of uncertainty – then fades and we attain a sense of finality, of realized purpose, through Ferraiolo's construct.

Ferraiolo's *Map of a Future War* and *The End of Capitalism* are the most overt politically oriented works examined in this book. In addition to the ideologies she challenges, her approach to design also challenges the status quo of many hypermedia productions. Interactive collages with unusual navigational principles, these titles present different types of experiences for viewers. While nothing can be physically added, the work possesses plasticity in its narrative formation and a customization of its serial reception. *Map of a Future War* particularly, perhaps uniquely, partakes in the Concrete tradition by assembling signs and symbols so as to create a tension between the two. Readers navigate the unexplained, ever-changing terrain of interface to find the same general senses communicated by (or emerging from within) disparate materials. While providing a wide range of routes through both works, Ferraiolo's message is steady. Her process involves sampling (appropriation of symbols and video clips) and mediated connections (intertextuality). Through these historically tested devices, accompanied by original written passages, she crafts variable traversal methods using contemporary modalities. Ferraiolo offers viewers a sense of freedom in deciding where and when to move, but as author/programmer directs the audience's experience. Her creative implementations, using the machinery of capital, i.e. computer technology, to point out flaws in the system's trappings, are not propagandistic but rather suggestive and critical as they expose the abstractions and suffering caused by economic injustices.

Mary Flanagan

[theHouse] (2006)[114]

Mary Flanagan is a professor in Film and Media Studies and the Sherman Fairchild Distinguished Professor in Digital Humanities at Dartmouth College. She is author of *Critical Play: Radical Game Design* (MIT, 2009) and has written extensively about human values in games and other 'technological, representational, linguistic and experiential' systems; her blog is titled *Tiltfactor: Game Design for Social Change*.[115]

In *[theHouse]* Flanagan uses a unique, graphically compelling (i.e. extremely stylized) approach to design an intimate, yet impersonal,[116] screen-based poetic environment. Through this mechanism, she, like Bouchardon, suggests the inevitability of chaos or a lack of control, even when circumstances suggest a certain amount of viewer regulation of the interface is possible. Viewers indeed have some control but might mistakenly believe they have more than they actually do because the virtual object – quite significantly – also capably moves on its own if viewers do not interact. Viewers negotiate 3D space on the screen, an effect offering a sense of moving through the galaxy of a poem, with a host of constellations and accompanying debris. Clusters of 3D cubes appearing in shifting grayscale tones spin and change size on their own, accompanying multiple, reverberating phrases on the screen, projecting the sense of a satellite rotating in space. Constant movement of this revolving aspect fluidly attracts the eye.[117] The poetry, somewhat copious in quantity, also changes in content and shape. Positioning the elements involves manipulation of textu(r)al information by the viewer as he or she absorbs the work. Passively drifting through *[theHouse]* is possible – the viewer can effortlessly watch an animated poem transpire on the screen – but such an approach is discouraged by the mechanisms presented by Flanagan.

Viewers quickly discover that clicking on the screen changes the content of the textual passages; clicking and dragging the cursor on the screen brings forth numerous alternative views of the objects and all content. Within *[theHouse]*, viewers most commonly see a pair of phrases appearing multiple times, sometimes in reverse, as if seen from behind. Brevity of textual information endows the phrases with a koan-like quality; in this respect they challenge the viewer to conceptually register and build a poetic context for them. Moving the cursor alters the position of the shapes and text; viewers move and pivot the cubes and words, which appear as free-form assemblages of phrases and boxes. Navigating through [theHouse] occurs on micro- and macro- realms on the interface. In a wider scale view (Diag. 4.2), symbols shaped and distorted in 3D dominate the view and offer an overall graphical perspective on the work:

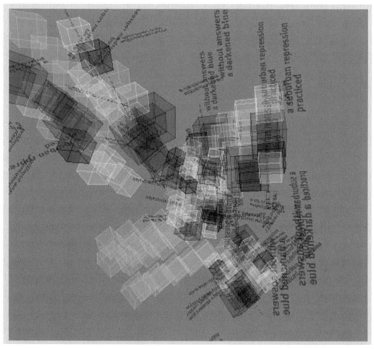

Diag. 4.2. Mary Flanagan. [theHouse].

Within a seemingly centrifugal presentation of language, the first lines to emerge in this particular reading, 'without answers/a darkened blue' and 'a suburban repression/practiced', describe a condition and a location for [theHouse] but otherwise lack intent and specificity.[118] At this moment the imbalance between text and image favours image, which in its multiplicity has a crowding effect. Viewers can certainly begin to ascertain the poem's scope from the text itself, as the colour scheme, scale and burdensomeness of the visuals also begin to establish thematic tone.

Moments later, with a few hand movements and perhaps a click or two on the screen, viewers become immersed in different texts and visual circumstances. With physical input, they traverse text: particular areas of the screen become magnified and appear to be drawn to the foreground and thus emphasized. Particular passages become privileged through heightened visibility, even if the phrases are even further fragmented, distorted or unreadable, effects apparent in Diag. 4.3:

Diag. 4.3. Mary Flanagan. [theHouse].

While difficult to read due to visual layering, a repeated verbal line (in multiple planes) dominates the screen: 'giving emptiness/never speak, for'.[119] In an otherwise silent experience, content of the text – partially overlapping and existing outside boxes that had previously been so dominant – does most of the talking. If a connection exists between proximity of the 'rooms' and text that appears, a viewer in this position might be led to believe that the closer she or he gets to a room, the emptier and more distorted it becomes. Looking at [theHouse] from a critical perspective, we see the centrality of fragmentation to the creative act; we can acknowledge the demands of putting the disparate pieces together in a pragmatic sense. These components may momentarily frustrate viewers, but also add vitality to the work. Flanagan's design demands an investment from readers; without bonding with the interface, and learning how it functions, the overall picture cannot be clarified.

A 'sine wave' configuration accentuates the succession of shapes ('rooms'), and imparts a sense of shifting geometry in both the lines and figures.[120] A flowing relationship, or oscillation between parts, exists in every aspect of the work. Progression and shifts between texts interconnect them, although not always directly. Flanagan sometimes maintains narrative continuity at the level of line; in certain instances, portions of a pair of lines permute. For example, the lines 'giving emptiness/tremendous', after a click on the screen become 'one thousand small/tremendous'.[121] As above, despite common language found, ambiguity persists in the permutation. Making vague connections between the digital structure, i.e. the boxes/rooms, and lines of the poem is possible, but the lines are not specific: viewers decipher and associate the lines, but meaning is suggested imprecisely. In the scenario projected in these passages, many small rooms may be the forces eliciting tremendous emptiness. In another sampling, lines 'library studies/you have a faithlessness' morph upon clicking into 'you have a faithlessness/unforgiving corners', then change course (after another click) with 'your hand floats on wall/without answers'.[122] Of course, no matter the order or style of presentation, the lines – as in any poem – compound one another, sometimes with direct repetition, which tends to instil an echoic, emphatic effect. The appearance of *parts* of phrases significantly contributes variation within the layers of repetition. For example, in one viewing I encounter at an angle the fragment 'blue', which disappears and returns accompanied by the word 'answers', all before the complete phrase 'a darkened blue/without answers' emerges.[123] Texts mainly interconnect as a result of how viewers choose to move and operate the mouse during the actual encounter; they may also accidentally move or be brought to a location within the structure automatically. In any event, some lines are fully revealed, and others not; meaning and message change as a result of the variant encounters made available by Flanagan's programmatic structure. Multiple appearances of each phrase also impart repetition, and accidental combinations of phrases, through overlap or montage, become part of the same visual field by inherent design. The centrality of permutation to Flanagan's work – a primary foundation for many digital poems – is obvious. Mutations of line, permutations of line and visual space at once enable a floating, shifting, rotating poem, a complex cluster of interconnected geometric expression.

Flanagan's description of *[theHouse]* informs viewers of its autobiographical character and explains that she enacts 'texts within, alongside and outside of the text of computational code'.[124] Her presentation, prepared using the open source program Processing, is 'written upon "rooms", and these rooms emerge to create "houses" next to and among the intermingling text.'[125] Lines embedded in the work clearly indicate disconnections resulting from claustrophobic conditions. Magisterially, Flanagan has found a way to reflect similar dynamics through the constant 'sine wave' movement's influence on the geometric patterning. Visual effects embody their own illustrations of detachment while simultaneously projecting actions mirroring those being pronounced in the poetry. Text and object fields technologically unite and thematically connect. Viewers can never reconcile the separate components; instead they build an understanding of them while navigating through the work, at first gaining perspective, then working to fuse coherence while struggling with the machine-modulated visual and textual differentials contained in the work.

Each aspect of *[theHouse]* adorns the work with aesthetic properties, some of which serve to interfere with its visual field. Locating readable text is not a problem; viewers can easily find and interact with verbal materials. Relocating text or specific structures within the layers of the work is, however, among the challenges (if not impossibilities) of the poem. In presenting difficulties in navigating through the piece, which mirrors the drama of *[theHouse]*'s metaphorical occupants, Flanagan achieves success. Given some, but not full, manual power over the work, a completely orchestrated ride through the materials is an impossible achievement. As phrases amalgamate and/or disappear, the transmission of direct content ceases. Watching and perhaps gently coaxing readable text to emerge gradually is possible, and minimizes the capacity of losing one's bearings in the poem. Doing so, however, represents a limited engagement, and does not account for and represent the full scope and purpose of the experience.

In *[theHouse]* we encounter shapes and lines that indicate and characterize the work as self-referential. Passages such as: 'holding rooms, very square/without answers', 'unforgiving corners/ unforgiving corners' and 'never speaking/never speak' seem to address the work directly.[126] In addition to the poetic ambiguities described above, once we turn a 'corner' in the work, through mouse movement, we may not find our way back to our previous verbal and visual perspective. Flanagan crafts lines so as to connect with the gestalt of the poem, which 'speaks' for itself through its texts and shapes. Her poetry comments on challenges emphasizing lack of fulfilment inherent in communication:

> giving emptiness
> letters have their sharpness;
>
> sometimes
> unforgiving corners.[127]

What are the unforgiving corners addressed by the poet/ programmer? Are they the edges of the interface that serve to cut messages off? These words, combined and recombined, can imply many meanings. 'Unforgiving corners' suggests the impossibility of reversing decisions, as well as the hardness of walls. All the verbal and visual interplays relate to relationships in the sense that words and language, our main source of communication, are accompanied by body movements and gestures. Corners and words can be sharp and unforgiving; one needs to take the time to think through decisions and ideas before taking even simple actions, and the more important an action or decision, the greater and more intolerant the consequences may be. Each of these potential reading issues factors into the drama unveiled while interacting with [theHouse].

One might expect something more fixed or organized from a work titled [theHouse]. In actuality, its interface design initially mystifies viewers with its 3D orientation. Other examples of literature that employ a building's physical structure as well as its secrets and tragedies to confound readers, such as Edgar Allan Poe's 'House of Usher', may come to mind. Throughout [theHouse], an inveterate emphasis on procedure, coincidence and participation dominates. These effects arouse intrigue and destabilize the viewer's experience, which viewers may read as a reflection of what happens inside the (now) metaphorical houses depicted by Flanagan, within which two people in a relationship come apart. Viewers struggle with text that represents the human struggle of the couple. This struggle, like that of the viewer, lies within the inability to locate (repair) themselves and the relationship within the exchange. In both frameworks, this perhaps occurs because of unanswered questions, deafening silence, and other presumptions about or expectations of the narrative. Through the text and words, viewers acutely sense the poem's atmosphere. A struggle between the viewer and mouse while attempting to read text and command graphics makes Flanagan's story come to life. Text does not always perform in ways the viewer directs it to, a conundrum relating to the couple's struggles within a relationship and a confined space. A visual trait of the work that supports this notion is the intertwining of Flanagan's houses; they seem to deny separation or the ability to escape one another. These houses can drift on their own, like life and relationships; their movements, while similar, are not identical: they can float around in disarray and come together to form an identifiable shape. The movement of words connected to parts of the house represents movement of people through their daily lives; where we build blocks depends and is dependent upon relationships formed through language. This may relate to the autobiographical couple's desire for space they never get. Yet the viewer also begins to feel crowded, and constantly changing text represents an argument in which people are talking over each other so only a small amount of dialogue is understood. A vivid portrayal of a relationship that goes out of control, [theHouse] considers the nature of companionship and leads viewers to draw a conclusion which will vary in delivery depending on how navigation progresses and the sequence in which words appear. Building blocks are the fundamental unit of any house, and must be used wisely to construct something purposeful, similar to the way in which words and phrases erect meaningful transmissions in the poem. However, Flanagan intimates that such a formation cannot achieve fixity.

Who has the upper hand here, author or viewer? Viewers partially control the literature and images, but Flanagan provides the verbal and symbolic programmatic content and thus remains in charge of the experience's general domain. Like *The End of Capitalism*, *[theHouse]* self-propels, *and* is interactive; these dynamic settings separate these works from most other digital poems, which either project animations or explorative objects, but not both. For the viewer who does not participate, words gradually devolve into a 3D geometric dance of symbols into which text eventually resurges. These periods have the effect of temporarily calming the stream of dialogue, though text always recurs. Viewers whose initial intrigue makes them seek a complete reading of the text will ultimately get a sense of exhausting its possibilities or will run out of patience because an overall general redundancy within the encounter occurs and the confines of the buildings (as well as the scope of the text) become evident. In fact, viewers may find themselves wanting to get out of *[theHouse]* just as the people in the relationship reflected in the work do: the relationship between viewer and poem becomes as frustrating – if not more frustrating – as the emotional relationship portrayed. Yet in spite of any limits presented due to its scale – which might account for a preference for short-term viewing – a sense of the work as a prototype for a more largely proportioned production can be imagined.

Worthy of note is that *[theHouse]* was produced using an open source (i.e. free) program (Processing) designed 'for people who want to program images, animation and interactions'; it's also noteworthy that Flanagan provides the source code for the work.[128] Someone who wants to make something with Flanagan's tool could do so if they knew how to manipulate the code properly. Predecessors to this work include Ladislao Pablo Györi's VRML demos ('Virtual Poems') and Rosenberg's *Intergrams*, but in *[theHouse]* movement of cubes and texts cannot be stopped or paused, and sometimes appear in reverse – circumstances interfering with convention and presenting a significant challenge for anyone.[129] Visual juxtaposition, kinetic structuring and partial lack of control complicate the peripatetic and rapid reading necessitated by the work's design. Negotiating and putting together the fragments contained within the structure impose difficulty on the work. Despite this condition, *[theHouse]* reorders itself with virtual beauty, emitting phrases in motion, inviting viewers to disturb and unveil. How could an author, working on or off the page in a previous era, accomplish the combinations of formulations Flanagan does? Someone could obviously turn the page of a book and receive a new set of lines or stanzas, but the interactive reconfiguration of visual elements – which contribute so much to the dynamics and meaning of the work – could not have been achieved previously. Partial (magnified) or quirky (inverted) formations of texts are features not new to poetry, but the intrinsic, fluid (in motion) association between verbal and visual object in Flanagan's poem adds new dimensions to the literary object.

Mary-Anne Breeze (Mez)

ID_Xor-cism (2006)[130]

Baldwin writes, 'poetics can be described as emissions and spews of ASCII and plaintext messages in the network as the degree zero of electronic poetry.'[131] While many progressive poets have during past decades used typography in compelling, inventive ways, no writer or digital artist yet has illustrated Baldwin's point more clearly than Australian artist Mez (Mary-Anne Breeze), who has since the mid-1990s used her own hybridized 'net speak' language, mezangelle, to convey themes and propel dialogue in her works.[132] Mez first used this language in MUDs and online chat channels; she continues to do so at present in an array of Web-based forums.[133] In mezangelle, Mez employs text and text format manipulation to give an organized set of letters multiple constructions and meaningful dimensions.[134] Mezangelle is replete with plasticity: using braces and square brackets, as well as other symbol and punctuation keys often found in programming code, she customarily divides and imagi-natively re-joins fragments of words. Mezangelle also plays with homophones and makes use of colour, particularly on the WWW, which facilitates use of colour; words emerge or form as a result of their sounds and colours. Mezangelle, seamlessly embedded, adds a layer of complication to reading. No longer can phrases presented be considered at face value; instead, words are spliced within words in a language borrowing from colloquial and computational speech.[135]

ID_Xor-cism, classified as 'Wurk 1 in the W[n]e[t]b.Wurks_ Series', is a projected work made in JavaScript and HTML featuring mezangelle and digitally processed imagery.[136] Once loaded on the screen it requires no further mechanical action for the viewer; it contains no hyperlinks. Here Mez stages textual materials in a linear composition that always appears identically; text and images, set in motion, slowly scroll from right to left over the course of fifty seconds. The eye easily follows its phrases and imagery, and viewers can scroll backwards through the work using a scrollbar. Further, Mez constructs the work so as to loop back automatically to its starting point, to begin again upon completion. These are useful design tactics, because the work requires multiple viewings for full absorption of its content.

As seen in Diag. 4.4, two narrow lines of text appear above and below a wider horizontal band, which contains images and typographic symbols; essentially no text appears in the middle row:[137]

Diag. 4.4. Mary-Anne Breeze. *ID_Xor-cism.*

Due to its scale and oddness of illustration, Mez's stylish, striking imagery – symmetrical, distorted body parts placed within geometric settings – becomes the primary focal point. Looking beyond the decorative qualities of the five succinct depictions – which are appropriate for the digital condition, setting and concerns of the piece and in fact mirror the mezangelle writing technique visually – these images and adjoining symbols anchor the narrative.[138] In terms of design, everything contributes logically and skilfully to the overall flow of the work. In pictures and glyphs, Mez visually knits together not only the disfigured faces, but also symbolic coding gestures and a type of chromosomal braiding most evident in the third and fourth images in the sequence, shown in Diag. 4.5. In the diagram at the right in Diag. 4.5, 'x' figures – separated by underscore marks in Diag. 4.4 – fuse together, and to their left, beneath the line reading 'DNAesqueness.echo+vision.dubble', an 'xy' configuration binds what materializes as a Cyclops-like eye in the construction.[139] These braids, ostensibly genetic stitches, appear in both diagrams in Diag. 4.5: one used to join the ruptured Cyclops head, in which one ear appears as if it is a foot, on the left, and the other as connective tissue in a more ambiguous structure, which may imply construction of two different kinds of eyeballs, or ways of seeing at the left and right in the image. The sequence's final image (not shown) materializes as if to represent a flag and an hourglass made of fleshy symmetries and symbolic aspects of computer coding. Unification occurs in *ID_Xor-cism*'s middle row through symbols that one might associate with mathematics and science. All these attributes contribute to a sense, or acknowledgment, of change in what humans are, how we express sensitivities, how we communicate, what and how we consume and what our literature can be.

Diag. 4.5. Mary-Anne Breeze. *ID_Xor-cism.*

Text associated with these images incorporates mezangelle, SMS language and conventional, i.e. less obstructed, speech. Passages appearing in *ID_Xor-cism*'s upper line feature a sans serif Arial font; with one exception, i.e. 'N.' between 'float.' and 'seethe.', letters (words) and periods are grey; brackets, pipes, plus signs (+) and the 'N.' are pink:

[ur.blather.is.my.concept.meat] /
[desire.blan.dish.ments.float.**N**.seethe.with_in.my.chest /
|| my.skull] /
[echos.of.the.past /
+DNAesqueness.echo+vision.dubble] /
[fugue.blind+project[ile]ing.luxuria].[140]

Although Mez does not use standard English, each line or phrase here can be understood if closely considered. Let's begin with a line that's integral to the concept of the piece: 'Your blather is my concept meat', which indicates that foolish talk serves as the foundation for what she makes.[141] Flattering speech of desire, perhaps from the past, 'float' and 'seethe' within the persona's body, causing some sort of genetic reverberation and abnormal physical responses to what the persona sees or goes through as a result of hearing implied verbiage. The final (mezangelle) passage in the top line, while not made too complicated via the projecting/projectile fusion, is more difficult to interpret. Read outside the context of the lower line (discussed below), it perhaps responds to the sense of haunted experience portrayed elsewhere in the passage, thereby reflecting on a scenario in which music (i.e. 'fugue') being sung does not see the world properly (i.e. is 'blind') and issue from a state of self-indulgent sexual desire. The uncommon word 'luxuria' could be understood non-literally as a state of being in luxury, which would be interpreted as a different kind of activity, perhaps as decadence.[142] Given the non-explicit imagery inscribed, viewers consider and balance its contexts with other materials presented. Poetic ambiguity and layering permits multiple possibilities to resonate.

The lower line of text shares the same colour scheme (grey/pink) above a pink background, with a different font (serif, Courier New), suggesting an alternative voice:

i dreamt of ur curling geo_edges city last nite /
very peculiar /
||[(k)nots in MyBook[read:**do**(A)**cumen.t**].|| /
[i fall + heart.wound easily] /
[sporadically] /
[ma**jest(er)**ically] /
4 all things simulcra.[143]

Two lines feature mezangelle; the others, more straightforward, conjunctly correspond with upper lines to keep the narrative afloat, although the last line here certainly encourages the likelihood of misreading as many viewers will perceive 'simulcra' (which is the name of an old video game) as simulacra. Mezangelle lines make already unordinary language more expansive, in essence through the use of different colours, typographic symbols within words and crafty slippages in word selection. These examples complicate and enliven Mez's expression in ways that those above do not. The first line, the knots/nots pairing, offers two different meanings but otherwise is not complex; it implies that something is either tied up or contains negatives/pessimism. 'MyBook' is unspecific but does appear below the upper line 'my.skull'; a connection between thought and product, which possibly refers to the present composition, is plausible.[144] The bracketed phrase possibly offers an explanation for what MyBook is; it is as if it says 'read as' this '**do**(A)**cumen.t.**' Here the words *document* and *acumen* (i.e. knowledge) are clearly present, as is, phonetically, 'redo'.[145] When Mez presents the bracketed neologism '[ma**jest(er)**ically]' in relation to '4 all things simulcra', the words majestically emerge, giving a pause for viewers to consider whether the utterance portends something lofty or merely jokes.[146] Verbal plasticity inherent in mezangelle enables many interpretations. Since most viewers will not discern the presence of the obscure Amiga/Atari game (1990), a type of inside joke issued by the author occurs, although the misreading (i.e. simulacra) also relates to the context of Mez's work and plausibly plays a legitimate role in the scope of the poem.[147] What do mezangelle and images in this work represent if not the artificial likenesses of bodies, cellular operations and language?

ID_Xor-cism, presented as a linear composition/formation, puts into motion different tracks of language, to be read on the fly.

This effect almost serves to mask alternative reading orders and a sense of pathos contained in the words. Nonetheless, Mez's prosody presents several reading possibilities or ways to receive the language of the work. Pairs of lines, presented together above and below the first, second and third images, portray a narrative sense when read top to bottom as each image scrolls past, telling a different story than when reading both lines as wholes. Again, Mez discovers a way to splice words and phrases within other words and phrases to significant effect. Alternating top to bottom offers a different account than does reading top lines first, and then bottom lines. For example, the succession '[ur.blather. is.my.concept.meat]/i dreamt of ur curling geo_edges city last nite' delivers a different context to begin the work than does '[ur. blather.is.my.concept.meat] /[desire.blan.dish.ments.float.**N**.seethe. with_in.my.chest': one presents the projection as a dreamscape, the other as primary feeling, onto which the dream is later added.[148] Such pairs of lines appear above and below each of the five images, but in the second half of the piece words appear interstitially, and would then be read (top to bottom):

[echos.of.the.past /
+DNAesqueness.echo+vision.dubble] /
[i fall + heart.wound easily] /
[sporadically] /
[fugue.blind+project[ile]ing.luxuria]
4 all things simulcra.[149]

Is it possible to reconcile Mez's effort here as the fetishization of a little-known, archaic video game? The line 'echos.of.the.past' does precede the reference, and perhaps the one game could represent nostalgia for an entire legion of primitive and now forgotten titles. Falling and having 'heart' 'wound easily' seems, however, a less likely response to a video game (unreferenced elsewhere in the work) than it does to Postmodern conditions of relationships, bodies and texts. A much more likely way to interpret this construction would be to connect the lines and conclusion through larger issues, such as simulation within cyberculture and graphical aspects and other themes already identified within *ID_Xor-cism*.[150] These, and possibly other legitimate approaches to interpreting content, cumulatively establish the range of meanings incorporated within the poem.

As Helen Burgess astutely writes in the Introduction to the 'SpaceWorks' issue of *Hyperrhiz* (2007), '*ID.xorcism* invokes a reformatted body that simultaneously hijacks the dead visual space of the browser window and the 'curling geo_edges' of sampled skin, reminiscent of the detached virtual flesh-ovoid we see in the old 'floating finger trick'.[151] In the foreground of Mez's piece is a meditation on how our bodies, human relationships and emotions (not to mention artworks) are visually distorted, evolve, or at least change – for better and worse – as a result of digital technology. Emotional, physical and intellectual coding(s) found within *ID_Xor-cism* respond less to mass and cyberculture than they detach, yet intimately reflect on, personal experience. A 2009 Twitter piece (*[twitte]reality_fiction: 27/3/09*), fusing previous considerations on network writing with new twitter posts, tells of Mez's process and overall interests; it ends with a charge or order to its readers: '+ now: pls make = real[time]ity_fictions [tweeted>(google)waved> asynchronously_poured] that echo[lalia] + hint @ bac-+4th-consciousness-grounded   constructions>P[sychE]ro[s]cesses.'[152]   Mez encourages a technological authorship that is unafraid of revealing one's intimate thoughts and corporeal experience filtered through media and inventive language.[153] Back and forth consciousness-grounded construction, which primarily confronts psyche and eros, pervades *ID_Xor-cism*, although its interpretation follows a necessarily more complicated process due to its compositional strategies and, to a lesser degree, its reference points. Mez uses the space inside her lines of poetry to stitch these subjects together through internal and exteriorized projections of something both imagined and lived.

Interaction in *ID_Xor-cism* involves interpretation, but a singular explanation does not avail itself. Is the 'ID' being exorcised or exercized? Is ID identification, or the psychoanalytic term for the source of impulse based on the pleasure principle, the id? Mez's tricks with language bring together unlike perspectives; viewers can see it both ways. As have many poets, Mez combines text and images. Her horizontally oriented, looping scroll enables motion and progression. Viewers must not only read (and sometimes interact) with the work but also translate the mezangelle language; this process creates multiple interpretative possibilities. While not the first poet to invent her own form of language, she is certainly among the first, along with Sondheim and Cayley, to adopt verbal constructions emerging from the digital era into literature. Mez has obvious interest in playing with and making as much as she possibly can out of ASCII code; the addition of images gives viewers more to embrace. One way of looking at the work is to see it as an expression of how alteration, confusion and types of ruin play a heightened role in our lives, conditions easy to reflect in texts utilizing digital media's heightened verbal and visual extensions that, apart from other qualities, invite excessive use of representative symbols. 'The web does matter, spins up and out to affect our General Linguistic Economy,' writes Memmott, 'Today, tomorrow we prefix and suffix every.thing.'[154] Mez works this possibility – with her application of language, and by using images as prefix and suffix – to extreme ends; by doing so she metaphorically hits two, three, or more notes with a single gathering of letters in ways previous generations of writers could not.

Jason Nelson

I made this. you play this. we are enemies (2008)[155]

As the WWW became a widespread public fixture, digital poets became inclined to engineer works that take the form of a game. In addition to productions by Andrews and other artists, Jason Nelson (an American teaching in Australia) has produced – among other experimental titles of digital literature – a trilogy of art games. *I made this. you play this. we are enemies*, built with Flash, is the second instalment.[156] Nelson has also engineered two related participatory, playful titles: *The Poetry Cube*, a tool that enables users to enter sixteen lines of permutable writing using a rotating 'cube', and *Birds Still Warm from Flying*, an interactive device based on the Rubik's cube that is reminiscent of Aya Karpinska's *the arrival of the beeBox*.[157]

At the outset Nelson's digital poetry game informs viewers/ players that the game's 'levels' are 'designed/destroyed' around the interfaces of Google, Yahoo, Disney and other websites.[158] Directions for play are simple: players use arrow keys to move, and the keyboard space bar to 'jump.' Other non-technical instructions offered by the author include: 'explore, explore' and 'stop trying to "get it"'.[159] Pursuing the work's experience trumps understanding its content. A mouseover layer labelled 'game?' appears at the upper left of the opening interface, which possibly is a response to foreseen questions as to the work's true identity: chaotic literary art or game? With regard to the gaming dimension, Nelson explains, 'There is a conflict, an Appalachian-style battle between the game maker and the game player, the artist and those wanky enough to like art, the poet and those that sing-song themselves through bittery selfish sexual whatnots'; he further comments on its purposes, writing '"figuring out" is for control-centred hedonists and sharks with bees for hair, such fast stinging chomping machines.'[160] Viewers could take offence at being positioned as such, but what Nelson really suggests through these imaginative postulations is for viewers to immerse themselves in its visceral experience rather than look for grand meaning.

In every level of the game, Nelson orchestrates visual bedlam through animated icons, other kinetic materials and a range of energetic, sometimes cacophonous sounds appropriate to gaming circumstances.[161] At times players hear the author's voice – particularly in statements signalling a mistake, e.g. 'you've purchased a farm' or 'reincarnate yourself', where their errors consequently return them to a level's beginning.[162] Such occurrences have the effect of re-exposing players to content seen previously, asking them to read it again if nothing else. The avatar in *I made this. you play this. we are enemies* takes form as a doctored life-saving ring to which appended hand-drawn lines add a 'bloodshot' effect; it is never permanently destroyed, so players can read unhurriedly while progressing through each le.'el.[163] As creator of the game, Nelson channels player movement; his game allows the avatar to stand idly as a player pauses or follows announced hyperlinks, e.g. pop-up videos. Contrived chaos hinders reading, a condition that lessens with familiarity; viewers can meander through the game without concern for other aspects of the artistic scenario. In contrast to *Arteroids*, consuming text and navigating interface need not occur simultaneously; these activities remain separate within the experience.

A close look at the first stage of *I made this. you play this. we are enemies*, titled 'Monopoland: Even this will someday die', indicates the overall compositional (if not partial thematic) strategies used by Nelson. He reconstructs Google Blog's interface as a gaming lever and surface for poetry. As indicated in Diag. 4.6, the avatar begins in the centre on the left. A maze of barriers must be navigated; to move vertically in the work, the avatar 'jumps' onto one of the two Google buttons that move up and down on the screen. Each time the avatar contacts one of the text hotspots, e.g. 'buy,' 'sell,' and 'soap', additional text and images appear.

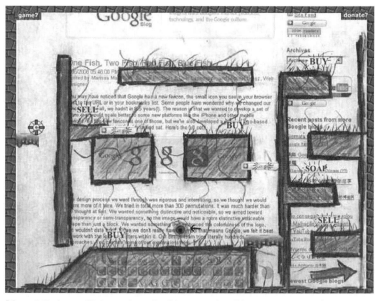

Diag. 4.6. Jason Nelson. *I made this. you play this. we are enemies.*

By the time the player reaches the final 'Sell' trigger in the labyrinth, and prepares to move on to level 2, the screen resembles Diag. 4.7. Traversing this interface, they also encounter several texts that appear and quickly vanish (emerging for five seconds or less). For instance, activating the first 'Buy' hotspot, an ephemeral text – a plaque emblazoned with the words 'simple is bleak hypnosis'

– flashes briefly across the screen, as does a star indicating the temporary availability of a video and the affixed text 'own all/rivers and/the cities/are yours'.[164] Since no alternative pathways exist, interactions with the game resonate identically for each viewer.

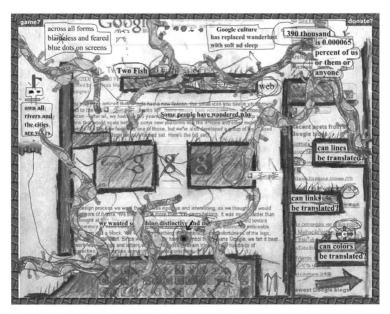

Diag. 4.7. Jason Nelson. *I made this. you play this. we are enemies.*

Scrawled pictures and other drawn elements, as hand-rendered gestures, not only decorate but symbolize the artist's usurpation and repurposing of corporate design and modelling. The carefully chosen Google Blog entry that Nelson defaces, titled 'One Fish, Two Fish, Red Fish, Blue Fish', introduces and discusses issues regarding the creation of a new 'favicon' (favourites icon). The entry discusses Google's aesthetic needs – which Nelson exploits when he highlights a line in the entry 'we wanted something distinctive and noticeable' – as well as their technical needs, i.e. to find something that 'would scale better to some new platforms'.[165] Nelson's literal utilization of the favicon as a vehicle upon which to transport the avatar handles the concepts of 'platforms' (horizontal surfaces) and 'scale' (to climb) completely differently: a creative literalizing not without humour or, within the game, purpose; wordplay involving the content of Google Blog directly influences his game's graphical design.

Texts overlaid in 'Monopoland' directly criticize Google, as in the lines, 'Google culture/has replaced wanderlust/with soft ad sleep', and also suggestively pass judgement, as in, 'own all/rivers and/the cities/are yours' and 'Some people have wondered why'.[166] The third 'Buy' hotspot activates the lines 'everything yours/is ours as you are' on the surface of a cartoonish trophy.[167] Nelson does not limit himself to questioning the power (and our use) of Google as he creatively builds poetry (and video vignettes) tangential to the company and its mechanism. More diplomatically, he asserts 'Everything will work together' inside a series of stars attached to another 'Buy' hotspot, and contemplatively asks, 'can lines/be translated?/can links/be translated?/can colours/be translated?'[168] Focusing on Google, while asserting his own power of design in conjunction with issues of WWW aesthetics, is a forceful combination. Nelson's playful additions capsize the relatively generic original projection of image and text, showing how someone can participate in and build from the WWW in unexpected ways, beyond consuming information or using a search engine. As discussed further in Ch. 5, Google now plays a major role in the distribution of literature *and* the crafting of literature and art on the WWW; that Nelson would want to begin his piece with Google as a reference point, with which to draw attention to critical issues, fits nicely as an opening.[169]

Much of the written poetry throughout *I made this. you play this. we are enemies* appears either within other graphical elements or as text cut-out from printed pages, featuring basic fonts in black and red, as in Diag. 4.7. On the few screens where fonts and presentation more typically embody ordinary WWW style, Nelson consistently implements his own hand and raw aesthetic to cover up the initial presentations. With the addition of more text and colours, verbal poetry becomes more challenging to locate and read. In levels 2 through 10, players experience similar navigational principles, albeit with different verbal and graphical material. Although techniques for presentation texts remain constant throughout the work, content (words) embedded into the remodelled appropriated interfaces, while striking similar thematic chords, differ at each level.[170] As levels pass, absurd, humorous, poetic passages appear alongside those easily read as critical. On level 4, for example, Nelson adorns YouTube's interface with the comical lines 'Muffin (cannibal)/of muffinville/ made some/muffins from/muffin meat'; on level 5 (Disney), viewers are informed, 'here comes/DEATH BY/DALMATION' immediately before the screen floods with cartoon dogs from all sides.[171] Midway through the game, Nelson stages an Intermission, offering the audience an invitation to '(go, go and have a constitutional and lie down)/(a few more levels to go)' while installing a video link and providing an email address and the instruction that players should write 'to say you made it this far'.[172]

I made this. you play this. we are enemies presents its viewers with a virtual obstacle course, at points overtaken by graphical features. Moments of self-reflexivity emerge, as in the 'Nauseotorium' (Disney) section (level 5). At this juncture, about halfway through the game, and immediately before bombarding the screen with dogs, the phrase 'pause to reconsider the joys and pains/of playing this game' appears, 'joys and pains' in red.[173] Nelson's poetry often asks us to consider what we do in the fields of information we seek, and uses poetry to invoke other realms of existence where we may also prosper. For instance level 2 (Yahoo!, 'Portalhoma'), as in level 1, invokes the humanistic, e.g. 'but we are/humans not/just bots/ and equations', then lines relevant to the business aspects of the WWW (and Yahoo!) blend with themes from the natural (offline) world: the poetic reminder 'wood is/fuel are/hits/are accounts/ are stock' precedes the lines, 'she needs stock/her fields lost/in the content/heavy rain likes' and 'the global/indifference/net held forest/balloons in trees/hot and tethered'.[174] Yahoo! is used to make an acrostic poem: 'Yes/Ask/Hordes/Open!/Open!'.[175] At level 2's conclusion, an exclamation, 'GROW!/DAMN YOU!/GROW!', adorned by smiling flowers, inundates the screen.[176] Nelson leaves us more than criticism throughout *I made this. you play this. we are enemies*, consistently fusing themes from within and without the WWW. At one point in level 3, the line 'knots as links' appears, in which Nelson symbolically bridges material and immaterial worlds.[177] Such purposeful connections serve to bring players to poetic modes of thought.

In level 8 (Metafilter, 'Melinkville: anonymous as fluoride'), after one set of phrases replace another, the screen becomes covered with words, many of which say 'text', while others say 'then other people say things', 'date and time and such' and, more poetically, 'cleaver notation'; the page converts to something more generic, thereby signalling the presentation of information as raw, empty words, to which viewers must always react in order to transcend to another level.[178] The final level, Joystiq, 'Gameration', clearly addresses the fact that this 'game' does not emulate a computer game in the literal sense. Nelson defines in the negative what the project is not. *I made this. you play this. we are enemies* is *not* about explosions, shooting, mitosis, corporality, fracturing, love, movies, time, drugs, games, teams or vice. As the avatar finishes the course, players face a polemical announcement regarding venues for the presentation of art: 'we do not need galleries any more'.[179] In lines somewhat faded in the background, Nelson divulges what his project consists of: 'some confusing/type of hell/built from/messy lines'.[180] *I made this. you play this. we are enemies* is that, and more. Nelson borrows iconography from the WWW, conducting an engaged experience as he re-contextualizes them with a humorous, purposeful, poetic and visually additive framework.

As players progress through multiple levels, running totals do not surface. Because this is not a game, Nelson instead jokingly provides the same score for every player at its endpoint. Upon completion of the maze, we are informed 'Your score is 42 and always has been,' and that 'a PLACE named THE END by YOU for NOW' has been reached.[181] *I made this. you play this. we are enemies* is not about the score one achieves. Its concerns highlight the creative imperative, and the need for people to be making something.

At the conclusion, Nelson offers three videos, which may not have profound meaning but unquestionably provide further evidence of his ability to perform creative acts; he not only says what should be done, but does it. Nelson's reverence for absurdity continues in each of the brief vignettes he narrates and manually acts out with objects in front of the camera. In 'Alluring ending video prize NO. 1', positioned as 'a history of the internet', Nelson manually animates a series of objects in a dish, concluding with a robot devouring a collection of hands removed from Barbie dolls.[182] In 'A cringe barking video winning NO. 2', Nelson uses an assemblage of household items to tell a story spontaneously. 'A hope breeding video gawder NO. 3' presents a fashion parade that, again, features use of body parts. Throughout this title, viewers negotiate and derive meaning through a series of digitally structured, mischievous, mix-minded poetic constructions, and may correctly detect a falsity, or non-seriousness, about the presentation – particularly since *I made this. you play this. we are enemies* doesn't seem to take itself seriously as a game. Nelson in his commentary and materials included in the composition says he does not intend to launch any sort of serious message through his work. Yet I believe he ends up doing so through his promotion of creativity. The work itself may or not be inspiring, but it sends a commendable and valiant message about the importance of being creative and using one's imagination.

I made this. you play this. we are enemies is quirky, provocative and lively in design and artistic energy. Unafraid of being unusual and unconventional, Nelson wants to be free of the responsibility to deliver a message; he aspires to entertain in a stimulating and confrontational way, as implied by the title. His pronouncement that viewers should 'stop trying to 'get it'' certainly contributes to this concept. Players of the game can recognize each level as a stanza and every aspect of the levels – graphics/sounds/animations/words – as components of the lines comprising the stanzas; finally, players can see that all the game levels in concert make the body of the poem. Playing the game, viewers negotiate a series of interfaces and encounter added texts, drawings, design elements and objects triggering media components. Nelson's objectives do not require reading for finite interpretation of content, but rather moving through materials in order to consider the experience of doing so. In the end we are left by ourselves considering on what foundations we want to move. Nelson's creation does not idly occupy time, but rather provokes us to say and to make something new out of materials at our online disposal. Reading or seeing everything Nelson presents in *I made this. you play this. we are enemies* is improbable; absorbing all its content is unlikely – a viewer would have to examine carefully and very methodically each stage, curtailing the urge to move swiftly through the work. Accumulating points does not influence interactivity and encourages unhurried progress, but even then viewers may not encounter everything Nelson inscribes. As in Rosenberg's *Diagrams* or Ferraiolo's *Map of a Future War*, incomplete readings are part of the poem's design and should not concern viewers, who can perceive its essence while partially traversing levels. The onslaught of information presented by Nelson precludes fully capturing the content. Nelson's poetry in *I made this. you play this. we are enemies* appropriates texts from exterior sources, fusing them with raw, open forms of original verbal and visual expression. Thus, Nelson uses and digitally expands appropriative techniques established by writers who have used cultural refuse as compositional elements in order to question the status quo and enliven human experience.

Stephanie Strickland

slippingglimpse (with Cynthia Lawson Jaramillo, Paul Ryan, 2006)[183]

Stephanie Strickland practises digital writing intensively: she has published numerous poems on the WWW.[184] Author of several printed volumes of poetry as well, her book *V: WaveSon.nets/Losing L'una* (Penguin 2002) contains a separate WWW component created with html and JavaScript. Entitled *Vniverse*, it is the premier example of an artistic work that illustrates the poetic possibilities for unifying a body of writing in traditional and experimental formats through computer technology.[185]

From a historical standpoint, the *slippingglimpse* collaboration can be classified as hypermedia, or a processed videopoem with hypertextual attributes. Encountering the work, made with Flash (incorporating digital video) and html, a viewer finds ten hyperlinked thumbnail images and a link to an introduction and makes a selection to begin.[186] In the introduction, Strickland and Jaramillo state their processes, offer basic instructions, and describe the project as a poem 'combining text with videos of ocean patterns seen along the Atlantic coast (chreods)'.[187] Since poetry has often traditionally aligned itself with nature, endeavouring to blend organic properties – particularly those associated with developmental pathways in space (chreods) – with digitally rendered (virtual) writing seems like a logical step for a poet to make at an historical moment when the possibility to do so exists. Entry points to *slippingglimpse* vary, with links displayed on a grid without hierarchy. Each thumbnail depicts a different view of water, such as a wave crashing or rippling surfaces; in one, floating leaves appear. This pied, colourful opening begins to reveal a thematic range of the videos, which portray kinetic aquatic movements that unceasingly return to a steady trajectory. Each link opens a videopoem in which processed kinetic imagery, i.e. videography produced by Paul Ryan, montages with poetic passages 'floating' in moving water. Words are mostly legible, although sometimes they are not very clear due to onscreen colour contrast. The cyborgian coupling of verbal and visual components does not become fixed by time and placement in the animated narrative; with each rendering of *slippingglimpse*, the various fragments of poetic text appear in different areas of the visual stage. Imagery on the video closely studies a depiction of water, in which image-capture video 'reads … reading for and enhancing water flow patterns'; the work's introduction explains how 'motion capture coding … assigns the text to locations of movement in the water'.[188] The gallant and inventive supposition here directly connects words of the poem with oceanic movements, fluidly weaving together images and text. In viewing the work, however, connection between the water's patterns and placement of words appears loosely if at all – a finite association between textual and tidal movements does not always come into view. Nonetheless, everything on the screen constantly moves. Even if the elements do not appear precisely in concert with one another, variable points of juxtaposition between language and image persist. This is not only an example of poetry in motion, but digital poetry reflecting nature in motion. Artificial, programmatic constructs display the powers of the natural world, which dictate the visual course of the poetry.

   Linking to the top row's second video, which depicts the surface tension of the sea, viewers encounter something akin to the following picture (Diag. 4.8), which, as all of the videos contained in *slippingglimpse*, offers options to access high resolution video, 'scroll text'[189] and regeneration beneath it:

Diag. 4.8. Stephanie Strickland, Cynthia Lawson Jaramillo, Paul Ryan. *slippingglimpse*.

Words and phrases appear in ornate grey, cursive letters on a background of water footage unique to this segment of *slippig-glimpse*, inevitably darkened but usually a blend of shifting light and dark. A shapely, open form poem emerges in this transcription of legible text:

                              into the body
          Laserglyph
    1 billion   years   turning back
                          evolution of notation
                                        hangs   onscreen
                              wait in
          generate   the work.[190]

Poetry derived in *slippingglimpse* involves 'sampling and recombining words of visual artists who describe their use of digital techniques'.[191] When amalgamated, in this instance, a self-reflective association between geographic history and processes used by the work's producers emerges, making the point that the visual basis of *slipping-glimpse*, while contemporary, is also eonian: Strickland, Jaramillo, and Ryan make something profoundly new from something that has existed for ages. A parallel truth emerges from these circumstances: that poetry, an age-old form, can viably embrace and embody fresh characteristics.

As processing alters the videographic information, different formations of imagery and poetic fragments perpetually appear, partially parallel to wave motion, as in Diag. 4.9:

Diag. 4.9. Stephanie Strickland, Cynthia Lawson Jaramillo, Paul Ryan. *slippingglimpse*.

Here much of the text is layered palimpsestically at the right of the screen; some (as above) exceeding boundaries of the video imagery. Viewers can clearly read the thematically appropriate phrases, '1 billion/6-d hypercube/turning', possibly referring to geological ages and theoretical renderings, but struggle – especially since contrast between the text and moving imagery is poor – to read other combinations of layered phrases, e.g. 'day after day/year/bring'.[192] The concept of the chreod continually plays a role, as the text implies movement and perpetual return to a constant state. Once the video completes its course, text continues to morph in size and move on the screen. In the 'rise' and 'fall' of text i.e. appearance and disappearance, fading out and layering, endless new combinations present themselves, potentially offering new turns and dimensions to the original text based on identifiable premises.

Each segment of *slippingglimpse* features a set number of phrases drawn from poems presented via the 'scroll text' link and issued as two distinct and spatially matched texts – inextricably connected and read as a single unit.[193] Tones of these appropriated written poems, which scroll in a window beneath high-resolution video footage, can be categorized as technical (left side) and philosophical (right side). As reflected in the example in Diag. 4.10,[194] their presence expands *slippingglimpse*'s content by availing and mixing in otherwise unseen texts from the original source. In this instance, none of the extremely minimal text seen within the video appears in the clearly presented version beneath it:

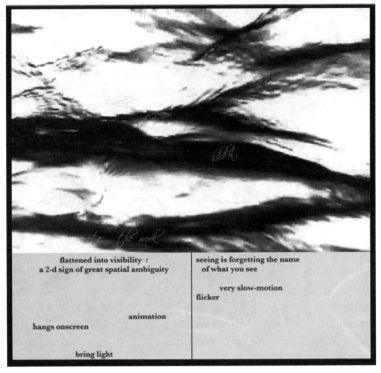

Diag. 4.10. Stephanie Strickland, Cynthia Lawson Jaramillo, Paul Ryan. *slippingglimpse*.

Textual content here reflects ephemeral natural movement as displayed in upper and lower fusions such as 'hold it/the work/flattened into visibility' (Diag. 4.10). The author writes on water, and asks the viewer to hold it – physically, a double impossibility, yet on a literal scale, one that a digital illusion upholds. We do need to register the text that 'hangs onscreen' in 'very slow motion' to grasp the poem's intentions, a possibility well within a viewer's range given the work's design.[195] In this title, processes of viewing are less demanding; we can see how Strickland recycles words, ideas, and verbal arrangements with others, and, simultaneously privy to all the source text, how permutation and mediation that treat language elicit recombinant energy and extend the initial statements or concepts.

Each section of *slippingglimpse* presents materials as described above. The artists deliver a consistent stream of messages and dynamics throughout the work. Unlike multi-staged works such as *Loss of Grasp* or *I made this. you play this. we are enemies*, each segment, while containing differing verbal and visual content, presents content uniformly. Strickland's efforts in general tend towards combining precisely what could be expected from a poet who turns to digital means to project expression: that a book would co-exist with a digital counterpart (e.g. *Vniverse*), and that nature and repossessed nature, combined with language given kinetic treatments, provide a valid direction for inventive artworks. Technologically-oriented beings cannot possibly ignore the powers of nature, their sway and the ultimate control (and beauty) they hold over us.

The absence of sound in *slippingglimpse* is curious. Since the ocean makes sound, as do words, an audio component (as an option, at least) would seem logical. But is this dimension needed? Are our memories keen enough for us to imagine sound and voices in the poem? This authorial choice is worthy of discussion, since many contemporary pieces contain sonic elements. Earlier in its history, digital poetry as a genre practically excluded sound. Once the WWW appeared, however, transmission of media hastened, and software enabling incorporation of sound became widely available. Not to include sound and design a 'silent' reading/viewing experience is a conscious choice. Historically, Strickland has shown preference for silent works; none of her digital poems contains sound. From a critical standpoint, the gesture must be read as resistance with purpose. Given the many fine qualities and innovations in *slippingglimpse*, Strickland's approach proves that sounds are not compulsory and should not be used gratuitously. Sonic elements potentially build in distraction from visual presentation and reading. All senses need not be stimulated at all times in digital poetry.

Strickland, Jaramillo and Ryan's *slippingglimpse*, an interactive multimedia art and poetry installation, uses sampling (collage) techniques to acquire and administer language. Exploiting the plasticity of the screen and animated letters, she creates a versatile, if insular, work that looks both at the exterior world and at the microcosm of possibilities encapsulated within a focused work that can be read, watched videographically, and interactively explored. Processes involved with its textual acquisition may not be new to literature, but the gestalt of the piece, and its elegance, offer a fine example of how digital literary works may reach unification with the physical world at large in striking ways.

5

Case Studies 3: poems of the Web, by the Web, for the Web

Thus far, most digital poems appearing on the WWW do not depend on the network but take advantage of it as a distribution device.[1] Several digital artists have, however, engineered interactive applications that harness information generated from transient mechanisms, such as Internet search engines and other WWW-based devices, to create hybrid forms of poetry. Net-art, a term used to describe artwork created specifically for the Internet, cannot mechanically operate offline. This chapter discusses the artistic values, successes and practices of artists who explore the expressive possibilities of net-poetry. Works introduced below differ from those explored previously in this book because they employ specificities inaccessible to offline media; they thoroughly rely on the WWW – without the network, these digital poems could not exist. More specifically, the case studies below focus on works that utilize common WWW resources, such as Google, to engineer digital poems.

As most of us know, Google's search engine is a hugely popular device used to find information; it is the most used search engine on the globe due to its branding, ease of use, and the quality of results it produces. Its predominance, visibility, utility and scale has led several digital poets to harness the search engine for art's sake.

However, before we explore this, it should be pointed out that net-poetry initiatives predate Google's rise in popularity. Text collages drawn from the Internet were pioneered in Mark Hansen and Ben Rubin's *Listening Post* (1999–2001), a multimedia art installation that compiled and synthesized texts from Internet chat rooms, bulletin boards and other online public forums for display and broadcast. Another early use of Internet text collage was *netomat (TM)* (1999), a 'meta-browser' installed at the Whitney Museum in which the artist Maciej Wisniewski invited viewers to type in different words; these words would trigger projected animations of network-associated imagery on the walls. In these artworks, information streams culled from the Internet mix and merge to create poems that feature ever-changing juxtapositions and intertextual relationships.

Although net-poetry predates the Google era, we will focus in this series of case studies on net-poetry created during the height of Google's popularity. Here is a brief overview of these studies: *Googlism* and the *Google Poem Generator*, like *Listening Post*, use network media to create reformed texts in WWW browsers; in other words media from the network is fed into a computer program to make literary (word-based) art.[2] Eugenio Tisselli's *Synonymovie*, Jim Andrews' *dbCinema* and Jody Zellen's *Without a Trace* use content from the network to render unique poems that contain visual and verbal information. Eric Sérandour's 'Tue-moi' takes a different, more conceptual approach that doesn't involve search mechanisms but nonetheless could not exist without the WWW. These experiments, while often imperfect (and in many ways rudimentary compared to *Listening Post*), intriguingly utilize the cybernetic structure of the WWW, thereby intimating a type of digital poem that will become more refined as time passes.

Googlism (2002)[3]
The Google Poem Generator (2003)[4]

Googlism as represented on the WWW uses JavaScript to retrieve information from Google to generate a straightforward litany (list) poem.[5] Users interact with the piece, which implements html and CSS for presentation, by typing a word or phrase into a text-entry box.[6] *Googlism* automates the arrangement of language: it regurgitates a slotted list in which the query string (the input word/phrase) precedes the verb 'is', followed by a snippet of text gleaned from the network. To illustrate the simplicity and imperfections of *Googlism*, here is an excerpt from a testimonial poem made using Stephanie Strickland's name as the input:

stephanie strickland is one of a handful of outstanding
stephanie strickland is an editor at slapering hol press she is a co
stephanie strickland is widely hailed as one of our most brilliant
    practitioners of and thinkers about cyberspace
stephanie strickland is the author of the prizewinning book
stephanie strickland is at it again
stephanie strickland is the author of true north
stephanie strickland is a poet and hypertext author
stephanie strickland is a print and hypermedia poet
stephanie strickland is our visiting writer for 2001
stephanie strickland is an adaptation – a reworking or re
stephanie strickland is not a person to ignore young people
stephanie strickland is the author of three books of poems
stephanie strickland is an award
stephanie strickland is a print and new media poet
stephanie strickland is the author of three poetry books true
    1999 boston review prize by heather mchugh.[7]

Googlism performs very clinically. The application itemizes results, returning varying numbers of (sometimes partial) lines depending on the subject's online prominence.[8] Primitive in concept and highly constrained, *Googlism* represents an early form of digital poetry automatically drawn directly from the network. Reading a *Googlism* poem online does not differ much from reading a list poem – a form with enduring historical presence – on the page. *Googlism* poems may feature a fragmented, even nonsensical, narrative, but they are certainly readable (and often create humorous verbal juxtapositions). Because the Google search engine analyzes language as scientific data and not for explicit content, a range of unintentional contexts may emerge in any given query, e.g. 'stephanie strickland is an award'. This aspect of its production offers to readers the possibility of surprise, and does not impede or alter how we receive the poem as an accumulation of composed language. Slotted generators, like those featured in Pedro Barbosa's *Syntext*, performed in a similar manner before the WWW existed, without the expansive database to which *Googlism* had access. Thus the WWW provides a more expansive, if less controllable, textual atmosphere for poets to work with.

Programmed in ASP, Leevi Lehto's *Google Poem Generator* was a sophisticated, if imperfect, application.[9] The primary generator on the site's homepage, after functioning sporadically for several years, was taken offline in April 2009 (one hopes for repair, but its removal could be permanent).[10] A secondary section of the project titled 'Patterns' consistently generated poems emulating various types of writing until its removal from the network. Patterns for which Lehto wrote programs included the poetic forms 'Sonnet', 'Villanelle', 'Couplets', 'Pantoum', 'Sestina,' 'New Sentence' and 'Collage'.[11] As in *Googlism*, a user's search string helped determine the poem's output; the input would be shaped into a poetic form through an automated process of language identification and selection.[12]

My involvement with *Google Poem Generator* primarily transpired while preparing animated poems and a multimedia Java (applet) project, '13 States of Malaysia', as a Fulbright Scholar in 2006.[13] A program employing the WWW as a database was unfamiliar territory that I sensed was important to explore.[14] I had studied generators extensively, but none drawing from such a vast information resource. Several offline programs had capably prepared convincing poems using succinct databases; could a reserve as enormous as the WWW be harnessed effectively? While exploring Malaysia and Malaysian culture, I experimented with the application, developing a multifaceted compositional approach to creating poems of place, object, event and experience, synthesizing my own original acrostic writings with edited output from *Google Poem Generator*.[15]

The following example – the final section of a poem titled 'Dragon Fruit' – illustrates the process and result of a 'Spenserian' sonnet pattern. First, the raw materials culled from Lehto's tool:

are three species of dragon fruit in
554.8K. previous / next – return
real dragon fruit as you can see in
overwhelmed by the dragon fruit, although

Hawaii). The plant. is native to southern
dragon fruit, foliage. – Dragon Fruit are
unique green dragon fruit, southern
out of hand. – However, there

of thing. Was kinda cool, but I am sure
fruit are distinct from prickly pears
Click on a Play title below for more
sparks for niche market farmers

– Fruit Database foliage and developing
farmers that are I have been working.[16]

Editing the text greatly reduced its content; about seventy percent of
the output was removed.[17] Original stanzas become:

                    in
Hawaii). The plant is southern

unique, green,
out of hand.

Was kinda cool, but I am sure

                more
sparks for niche market farmers

                    developing
farmers that   have been working.[18]

Poems made with this application unquestionably reflected the
technological environment from which they materialized, sometimes
in interesting ways. When asyntactical utterances are removed, a

collage Dadaist poem centred on the subject emerges. Changes in the example above, including removal of numbers, screen navigation references, and needless repetition, bring forth vibrancies of the subject in provocative and colloquial ways (e.g. 'green,/out of hand', 'Was kinda cool …'). By choosing which content to privilege, I could reiterate the agricultural focus, thereby celebrating the means of production (the work of farmers) introduced in a previous stanza ('where they are known as/Fruit foliage and developing/farmers have been working'). What begins as a type of scientific list poem becomes a geography lesson, then morphs into a down-to-earth observation on the varied network identity of the product; thus the finished work offers an integrated, broadly informed perception about the subject's place in society.[19]

Disruptive fragments, often full of technical clutter (bits of http addresses, dates, numbers and other sorts of technical information) denote poems made by the *Google Poem Generator*. These characteristics require more attentiveness on the part of the reader, who must perform a new type of filtering to bring forth meaningful elements within the generated document, even if the encounter strictly involves text. Even without the interference of graphical images (none are included in *Google Poem Generator*'s output), the unconventional, multi-rooted syntax demands atypical efforts. The absence, or distinct interruption and truncation, of traditional narrative processes complicate reading these verses – the language of Lehto's sonnets does not lyrically flow as one might expect. One learns to disregard the debris or alphanumeric noise and read what conventional narration allows, or to regard and accept this debris and noise as intrinsic to this type of poetry.[20]

I became infatuated with the *Google Poem Generator* and allowed it to do some of my work because it brought forth dimensions of stated topics I could not discover or learn about through my own eyes, senses, experience and research. This tool also fascinated me because the program provides information and structures at once expanded (by gathering content from the WWW) and distilled (trimming the content gathered into poetic form). In certain respects, the responsibility of a search engine is to find, sort and create a hierarchy of information, tasks enabled and compounded by the functionality of Lehto's device.[21] Given the complexity of operations involved, the result of the project, as ephemeral as it may have been, amazed me. Despite any perceptible imperfections, *Google Poem Generator* proved that the scale of a generator's database could transcend the dictionary and encyclopedia to include any sort of information. *Google Poem Generator* will inevitably pave the way for more expansive generative tools when literary-oriented programmers triumph over complexity and cultivate elegant methods with which to harness the network.

Jim Andrews

dbCinema (2007)[22]

Described by Andrews as a 'graphic synthesizer' and 'langu(im) age processor', *dbCinema* – db here denoting database – uses results from WWW searches to render a processed animated poem according to parameters established by the viewer.[23] As in Tisselli's *Synonymovie* (discussed below), Andrews' mechanism uses Director to assemble images and text after viewers have entered a 'concept' and other parameters.[24] Andrews has made two versions of *dbCinema*: a rudimentary version (0.50) and a formerly

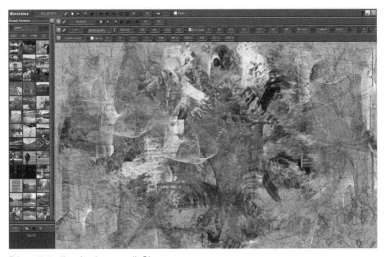

Diag. 5.1. Jim Andrews. *dbCinema*.

offline advanced version that now functions on the WWW.[25] Both use Google Image search and alternatively permit viewers to specify local directories of images, thereby making it possible to limn beautiful animations from both private and public sources.

In its ultimate form, *dbCinema* is wildly versatile, elaborately participatory, and produces high quality patterned art on a consistent basis.

dbCinema's animations actively flourish. Compiled images perpetually emerge, simultaneously preserved and layered on the screen. Andrews gives us a refined stage where viewers use 'brushes' to variably adjust minute details, such as the speed of geometry or the background colour, and establish important parameters. The 'brushes' devised by Andrews inscribe profound kinetic elements into *dbCinema* projections, enabling viewers to create and configure what happens on the screen. Viewers use his tool to 'paint' the 'movie' that they collaboratively produce with the network.[26]

Examples of output openly exhibit geometric designs, symbols, images and text.[27] Graphical aspects are dominant, but the subtle appearance of text draws attention to discernible verbal content. Through network location and fortune of programming, this verbal content intrinsically corresponds to the uniquely generated setting, as seen in Diag. 5.2:

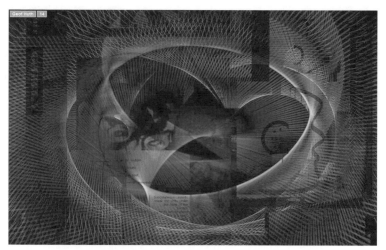

Diag. 5.2. Jim Andrews. *dbCinema.*

Andrews regards the program as a 'synthesizer', an instrument that gradually produces and unveils visual samples derived from a common location or based on a common theme. In Diag. 5.2, the clearly readable text fragments (which originated in Geof Huth's *dbqp: visualizing poetics* blog) offer an apt condensation of *dbCinema's* results:

multiple
past
immediate
Subcutaneous inserts
sneak the deity under
the bloat of the
swollen
to be reinstated (dbCinema).[28]

The many images already stored on Google instantaneously become a new work of art, one that subtly corrals iconography precisely for such a purpose. Any original qualities contained in the graphics become transformed into something unforeseen, particularly since the contents of the database may change at any moment due to the transient existence of information on the WWW. *dbCinema* draws instantly from a potentially enormous (or, possibly, a highly constrained) database of images and cyclically introduces them onscreen using a number of techniques. Because the program first veils images (foreground colours are established with the 'Stage Colour' setting in the Brush Sets menu), selected images appear beneath a cloak of colour according to parameters set by the viewer.[29]

Viewers can use singularly or simultaneously four 'editable' brushes – entitled 'Vector', 'Voctor', 'View Geometry' and 'Flash1SWF' – devised by Andrews; properties of each can be customized by the user to create different types of onscreen movement as images appear in the database.[30] The brushes offer viewers a wide array of options for orchestrating geometric motion and shaping forms: users are given 29 shaping tools and ten geometries to choose from. Each brush is assigned a concept (search string), and the images retrieved 'are used as 'paint' that the 'brush' 'paints' with'.[31] These adjustable patterns guide the positioning and overall appearance of the network pictures captured by *dbCinema's* 'thumb browser' (left panel, Diag. 5.1). Vector and Voctor brushes embed words or phrases into images, thereby giving users the ability to choose a font (four choices) and indicate scale, height, stroke width, seconds per word and dynamic opacity settings. For example, I embedded the phrase 'this is a picture about poetry' into the series of images contained in Diag. 5.1. 'Each brush', writes Andrews, 'has a configurable geometry, which is the path *dbCinema* uses to move the animation around the screen.'[32] The user chooses not only the images included in the movie but also determines its shape, motion and textual content. Language can be added in two ways: by privileging collected images containing words and letters, and by using the Vector or Voctor brushes. Text collages appear within the images, with picturesque interludes, accentuated by verbal qualities made with the brush. By actively conducting the images gathered in the 'thumb browser', users customize at the source of rendered output. Although optional, pruning pictures with unwanted scenes or colours can benefit the final result; however, this strategy requires the user to consume and response to thumbnail images with little delay.

When working with *dbCinema,* viewers must develop an understanding of what the brushes do and their aesthetic parameters, and manipulate the program's settings and the database delivered by the search mechanism. Given the number of variables ultimately available, optimizing the program's settings requires a considerable amount of exploration, patience and persistence. On the 'Brush Sets' menu, some three dozen settings are available with which to plot and produce animations. Minute details, such as starting points for x and y axis geometries, speed of rotation, or style and size of font, can be controlled through input boxes and pull-down menus. Less developed 'Image Settings' menus provide users with tools to specify important information, such as image size (both in dimensions and data filesize), WWW source, colour and content filter. *dbCinema*, more than any other program, gives its users copious opportunities to adjust texts emerging onscreen.

From amassed acquired images, users select any number of pictures to be rendered in a *dbCinema* movie. Without customization, readers will not find language in every frame or example; in fact, it's fair to say that without customization *dbCinema* will probably create something more painterly than literary. Nonetheless, Andrews has succeeded in producing a viewer-conducted, generative artwork that possesses a flexible relation with language. By using graphics as 'paint', giving each brush a concept, and enabling the presentation of language, Andrews' tool is a major accomplishment. Its output of compiled, processed, palimpsest images and texts can be very subtle – at any moment these elements can become imperceptible – but viewed cumulatively, across hours, the results are exquisite. In *dbCinema*, spirographic geometries create a net upon which images, symbols and language from the network entwine.

Eric Sérandour

'Tue-moi' (Kill-me, 1998)[33]

Throughout much of the history of digital poetry, authors have occasionally used the screen not to enthral the reader through an enticing intermediated narrative but rather as a device to make a poetic point.[34] Eric Sérandour's 'Tue-moi' is an example of such a construct that does so using the tools and language/protocols of the network. By design, when viewers access 'Tue-Moi' they encounter a blank white screen, which indicates what the piece contains – at least on the surface. However, upon closer inspection, viewers will notice a function that allows them to scroll down the screen. Doing so brings the viewer to a single diagram containing a collage of words (phrases) in French. These words appear to be identical each time 'Tue-moi' loads.

'Tue-moi' obviously incorporates visual components, and, upon engagement, reveals interactive traits. Amid cascading text and buttons, a rectangular box forms, containing static and mostly unresponsive elements. Inside the box, words and inactive buttons are layered, reversed and more textural than literal – a feature of many digital visual poems. This parcel of words and buttons contains no links, although the black bars do. A link on the right, when activated, simply centres the poem on the screen (as it appears in Diag. 5.3) via an html (#) anchor; the left link brings forth an untitled

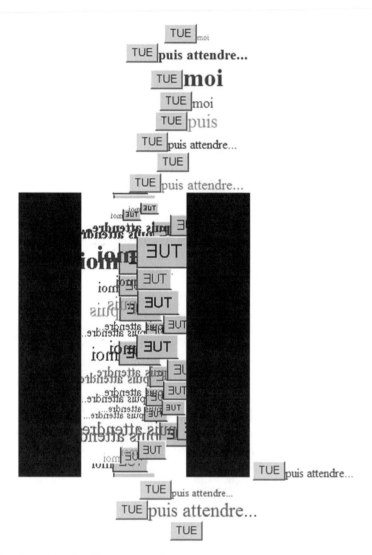

Diag. 5.3. Eric Sérandour. 'Tue-moi'.

page (noir.htm), on which nothing but a black surface appears. Looking at the appropriate section of source code reveals little about what one should expect to see beyond the colour of the pages, i.e. #000000:

```
<META NAME="Generator" CONTENT="Stone's WebWriter
    2">
<!-- Minus AutoDato -->
</head>
<body bgcolour="#000000">.35
```

If anything, a user without knowledge of html could be more confused by looking at the code, and might mistakenly believe some sort of generator should be present when in fact the <meta name> tag simply states the name of the html editor used to create the page. An experienced html programmer will know that an exclamation point inside an html tag (e.g., '<!-- Minus AutoDato -->') means that nothing on that line will show, but the novice will be puzzled and might believe something is missing due to the 'Minus' designation.

Above and below the box, twelve lines of text – each featuring a button labeled 'Tue' ('Kill' in French) followed by the word 'moi' ('me') or 'puis attendre' ('then wait") – appear in different colours and sizes without obstruction. Each of these buttons, in theory, performs the same task: opening up an electronic mail (email) editor, through which users can send a message to the address tue-moi@entropie.org containing the subject line 'tue-moi'.[36] The user might presume this designation is an email account managed by the author, but this would be unwise given conceptual possibilities of the digital poem.

In this reading of the work, the first disruption occurs in the form of a warning message that appears in the browser (Internet Explorer 8) after pressing one of the 'Tue' buttons:

This form is being submitted using email. Submitting this form will reveal your email to the recipient, and will send the form data without encrypting it for privacy. You may continue or cancel this submission.[37]

This message might completely halt someone concerned with online privacy and the security of computer data. While featuring a minimal amount of interactivity, comparatively speaking, part of the exchange with 'Tue-moi' involves engagement with software components outside the browser, as indicated in the message above. The browser routinely issues notifications such as this, a complication compounded in works involving more than one program. Warnings like these tend to create hesitation, if not resistance, from viewers. Technology employed here understandably leads to insecurity. Further, if an external email editor (such as Microsoft Outlook) is configured incorrectly on the user's PC, further warnings or outright failures will ensue, including periods of waiting around for something to happen. A litany of troubles may subsequently arise: over the course of reviewing this title on Explorer and Firefox browsers, I encountered four different error messages or warnings, and a message was not transmitted to the email address provided even when an email editor opened.

Some viewers will simply disregard Sérandour's piece as a 'broken' poem, especially those familiar with his animated work, e.g. 'Ceci est un mouche'. This assessment lacks the level of diligence required in reading digital works. At first I considered the possibility that the work's age contributed to some of the difficulties described above (it was created in 1998). Certain tasks once enabled by browsers may no longer function because the way browsers read and process information has changed, as have other software and network processes.[38] Finally, however, an exchange between the user and 'tue-moi@entropie.org' does not occur. Discontinuity in the piece might seem to stem from the user's PC configuration, but in fact the author has programmed the interactive mechanism's failure. Independently testing the address using my own email account in 2009, I received an 'Undeliverable mail' message, i.e. my test messaged bounced. The address was never valid.[39] Sérandour's 'mailto:' link offers nothing to users – except for the potential to spend time investigating its viability and contemplating what its failure to operate means. In the end, Sérandour has cleverly announced the principles of this work in the title and small phrases he chooses for the body of the poem: 'Kill me' and 'and wait.' Normally, the patience required for reading digital poetry delivers abundant content; true to form, Sérandour's mechanism delivers something thoughtful and worthy of contemplation; however he uses vastly atypical means to do so.

When users become fully absorbed in 'Tue-moi', they must resist the inclination to avoid a seemingly flawed structure. The work's design, however broken it may appear, is calculated; and even if it were not, it can be read as a textual performance or event. When we reach the end of the work (and an end *is* reached), we should consider the result of the encounter. Presumably, curiosity will drive the user to investigate and comprehend this creative expression via new mechanisms. They may ask: so what is the point of this work? To answer this question, 'Tue-moi' is a commentary on the failure of the WWW to sustain interactivity and reliability. 'Tue-moi' invites interactivity but does not effectively deliver it through electronic mail exchange. Conceivably, in a maintained, straightforward version of the piece, Sérandour could periodically respond to those who send messages in a variety of ways. An automated response could be issued, containing a generic reply to those who made the effort to engage the work fully. Manual or automated replies containing texts or images, etc. could be issued, thereby providing users with materials they may further consider and possibly use for their own benefit; alternatively the automated response may invite further engagement or exchange. Sérandour's responses could be offered as inspiration, art supplies, or as a sort of heuristic device. As it is, however, the work ends up containing a type of reflexivity in which exploration and exchange are stultified; rather than present the user with an external spectacle or an opportunity for one-to-one dialogue, it generates a one-way message.

Users confront fear of exposure followed by the risk of rejection: a message sent may not be read or replied to. Finally the user discovers that the entire process has been a ruse. The poem communicates its message indirectly and partially through the technology; its implications stem from its concept. Users confront a circumstance to which they must uncomfortably and ultimately unfruitfully (in terms of textual result) respond. In this manner 'Tue-moi' articulates an idea about processes of communication which are fallible and not necessarily realistic, even if technologically possible. When people read a book, they never receive a message that a page does not work.

In great contrast to this fixity, instabilities abound in Sérandour's work. Yet a perceptive return or capability to 'read' is, of course, still possible. In the interactivity, users end up directing or choreographing an empty experience. In the conglomeration of awkward visual and interactive effects, Sérandour uses the network to suggest links that take us nowhere and do nothing; he suggests that an author working in digital forms can capably score textual elements inside a super-ficially versatile box, flip them over, and deaden an experience.[40] On the surface, Sérandour's poem possesses many parallels to art: composing words and visual fragments on colour fields superficially corresponds to other artworks. However, the underlying premise of his work – in other words, the results of reading it – delivers to readers an utterly strange and different type of participation, one where transformative aspects lead readers into a dead end. Electronic reading here is not a pointless pursuit, but it has become an emptying, unfulfilling, disconnected experience, achieved as a result of its existence on the WWW.

Eugenio Tisselli

Synonymovie (2004)[41]

Eugenio Tisselli has written computer programs since childhood. At present, he teaches university courses in Barcelona while finishing a PhD at Z-Node, University of Applied Arts in Zurich. Tisselli's accom-plishments include authoring the multimedia program MIDIPoet, in which a user's sonic or musical input simultaneously generates images and texts. An epigraph once featured at top of his homepage, a Vilém Flusser quotation, explicitly reveals the underpinnings of Tisselli's creative orientation and practice: 'The easiest way to imagine the future of writing … is to imagine culture as a gigantic transcoder from text into image. It will be a sort of black box that has texts for input and images for output.'[42] In projects such as *Dadanewsfeed* and *Synonymovie,* Tisselli particularly succeeds in his quest to transform and translate graphically a user's written/textual input.

By adding media elements into the configuration of a net-generated digital poem, Tisselli's work heightens circumstances and complexities for reading. *Synonymovie* draws texts and images from the WWW to make a web-collage – a visual poem – comprised of image and caption. The program contains a simple but integral participatory element: users type a word to instigate the appearance of a series of associated words, paired on the screen with related images, i.e. images with tags in common with the generated words. To begin, users enter a theme into a text box. Next, by clicking on the start button, users activate a poetic chain of events. Both single words and short phrases are permitted; the choice of input directly influences the output, as one might expect.[43] Using a small PHP script,[44] *Synonymovie* retrieves images from Google image search and synonyms (sometimes also antonyms, according to the author's notes) from thesaurus.com.[45] Each time a new image/word pair appears, Lingo sends a query to the PHP script; the script makes its queries, and then returns results resembling output seen in Diag. 5.4. The 'movie' ends when no synonyms can be identified for the current word, at which point users can click the 'new movie' button and begin again.[46]

drain

<<< new movie

Diag. 5.4. Eugenio Tisselli. *Synonymovie.*

If users enter an uncommon theme word, *Synonymovie* typically (but not always) produces either a single screen of output or none at all.[47] Curiously, entering the word 'arf', for example, usually elicits a lengthy chain of image and text, none of which apart from the input word are associated with dogs. We can see this disconnect in the following transcription (the description of the image appears in brackets):

arf [road following riverbend in valley]
rind [cows in barn]
peel [two men reclining in chairs]
peeling [group of people sitting in chairs on a stage]
shell [image of Shell gas station sign]
pod [nude woman]
sheathing [a woodshop]
lamina [snowboarder]
sheathing [woodshop]
membrane [diagram of magnified skin cells]
lamina [console with knobs]
membrane [diagram of magnified skin cells]
film [a bicycle parade]
blur [shot of rock band]
make hazy [pastoral field]
assemble [pastoral field (same as above)]
summon [three men standing in front of tower, arms crossed]
mobilize [poster with African children]
prepare [emergency rescue workers]
fit out [image indecipherable]
preferable [children at bulletin board]
choice [woman posing in front of graffiti]
uncommon [collage of extreme sports]
aberrant [large snake]
deviant [Muslim girl with machine gun]
untypical [campground]
heteroclite [eponymous storefront]
extraordinary [ferry, mountain lake]
exceptional [image indecipherable]
unprecedented [distorted US flag, figure]

unrivaled [bear]
matchless [motorcycle with elegant sidecar]
supreme [men modeling raincoats]
absolute [abstract pixilated shapes]
absolute – the end [fighting fitness advertisement].[48]

With a little analysis, it is not difficult to follow how the program propels language in the poem most of the time. Nonetheless, how (or why) does this *Synonymovie* proceed from 'arf' to 'rind' or from 'fit out' to 'preferable'? Non-cognitive, mechanical decisions such as these challenge explicability.[49] I make these observations not to recognize the illogical jumps as flaws but rather to observe that these unpredictable progressions promote intrigue on a grammatical level. As language appears, in approximately three second increments, so do the synonymic, or harmonic, images. A relationship between verbal and visual elements is sometimes directly comprehensible, sometimes playfully (surprisingly) disconnected yet associable, and sometimes discordant (dissonant). Occasionally the imagery is completely unidentifiable, i.e. abstract and blurry. In lasting instances of *Synonymovie*, viewers must process visual information quickly while simultaneously considering – registering, tracing – the effusive accumulation of text.

From a literary critical viewpoint, attempts to refine the lyric are notably absent in the program's output. Whatever the network (thesaurus.com) gives, *Synonymovie* projects. In this respect, when compared to the poet who agonizes over word selection, often making selections based on how words sound, *Synonymovie* fails. Of course, the program does not aspire to compete on traditional registers; rather, it represents a radical contrast to the type of unitary lyricism evident in traditional forms of written poetry. Tisselli and many other digital poets do not aspire to reify lofty historical norms. Instead they employ different sorts of patterns, wherein programmatic randomness and machine cognition combine to synthesize network/media resources into a digital event almost guaranteed to contain turbulence. Readers may intuitively acclimatize to fragmentation and the absence of conventional syntax, traits not foreign to modernist and experimental poetry in the past century. The discursive narrative produced by *Synonymovie* subverts logic and the possibility of presenting an ordinary list of words, thereby endowing results with poetic tone. We read these assemblages as open form poems with stunted grammar and become accustomed to the work's quirky dynamics. When we do so, the initial turbulence diminishes – not necessarily because noise gets tuned out, but because our intimacy with the tool renders discrepancies and synthesizing fragments less formidable.

In Tisselli's scheme for *Synonymovie*, length of output depends on the number of synonyms associated with input words, and those generated subsequently. Serial synonymic connections populate the work, meaning that extended strings of synonyms bring more fruitful (i.e. longer lasting) movies. Results vary in content from rendering to rendering, even when using the same input; this is not only a positive aesthetic attribute but proves exactly how ephemeral net-poetry can be. While repeated input words often produce the same initial images, the thesaurus branches out and diversifies movies by delivering and distributing different synonyms associated with new images.[50] Commonalities between words in English do not necessarily lead to a monochromatic, sensible narrative; rather, the broad and arbitrary methods used to tag each image often promote

Diag. 5.5. Eugenio Tisselli. *Synonymovie*.

variance. For example, a movie with the theme 'river' included among its dozens of frames a screen that paired the word 'causatum' with a picture of a cat and walrus (Diag. 5.5), and concluded with the word 'hairbrush' matched with a likeness of that object.

In this manner the programming generates and constructs intriguing and peculiar movies. By modality, we receive them in a linear sequence, but the verbal and visual information presented – constructed 'on the fly' – peripatetically wanders; it takes unexpected turns as the program advances lines based on word associations. A viewer's ability to make direct connections from line to line and between line and image fluctuates, which in effect removes the work's linear sensibility; the perception of linearity also diminishes because every movie lasting more than one screen differs and is unique.

The inconsistent, sometimes poor quality of the visual images is a notable aspect of *Synonymovie*. Tisselli makes no effort to preserve the integrity of the digital pictures, which often appear greatly distorted because the program draws thumbnail images from an image bank and stretches them. This characteristic is integral to the work; it graphically indicates what each poem as a whole does: guides a fuzzy journey. Ultimately, one does not have a choice about what one sees in a *Synonymovie*; users may only initiate the threading of the program's textual weave. Because the program draws from WWW databases, Tisselli writes, 'every synonymovie is a movie about the net.'[51] Although the contents of a movie take on subjective meaning dependent on the viewer, the logical and illogical connections of the movies address and reflect the mechanics of the WWW. It is also important to note that abstraction plays an extreme role in *Synonymovie*. Abstraction is a concept not usually associated with the WWW, a tool that users typically access to find specific information and communicate directly without inter-ference. But as anyone who conducts research of any sort on the WWW quickly learns, even when doing a search on someone's name, the signal to noise ratio – i.e. the time it takes to find relevant information – is often fraught with diversion and dead ends. Tisselli capitalizes on this disjunctive quantity of the network by reducing its vastness while orchestrating a textual experience that undermines its facade.

Synonymovie requires from readers a willingness to be patient and to look past any need for perfection. Spending time with the program studying what it does and what works best, i.e. experimenting with different input values, may produce surprising results. Viewers can experiment by entering input intuitively (spontaneously) or strategically (consciously). Either way, it takes time for viewers to develop the type of familiarity and trust in the application/poetic instrument that brings good results.

Synonymovie is a vaguely determinate *dérive* (both in the sense of drift and of tabulation) through the body of an online dictionary synchronized with images on the network. Screens pass quickly but not before viewers register words and pictures. Connections between text and image can be difficult to discern, but these distorted configurations communicate and contain their own type of elegance. Because semantic connections between words and pictures resist blatant identification, the blurry pixilated presentation of images is appropriate. *Synonymovie* not only 'plays' with material aspects of the images but with what they mean. Contrasts between image and text require viewers to make an individual connection; thus reading and viewing become a highly subjective exercise. While watching a movie, viewers take a mental excursion that often requires them to make connections quickly, often intuitively and illogically. No predetermined end result is achieved apart from 'the end' imposed by the program after it runs out of connections to make. Viewers whose movies end too abruptly may experience dissatisfaction with their destination, expecting more. Others will find *Synonymovie*'s cognitive demands jarring. Since animations created in *Synonymovie* are ephemeral and unique, every user should experiment with the work repeatedly to receive a range of different transmissions.[52]

Conclusions derived from *Synonymovie* may be a statement about the network, about its aesthetic values and/or about the nature of associations between words and images in a network-generated text. Beyond these potential outcomes, more qualities exist. At very least, the emerging open-form, generated poems make a statement about language: the separation between seemingly distinct words and concepts may not be as wide as they seem. The power of antonyms to direct and guide a narrative and create a type of dialectic also becomes evident in poems that extend beyond a handful of screens. Engineering an online dictionary and search engines together is a significant feat, and Tisselli's exercise succeeds precisely because the combination of word and image in the program's output operates on multiple levels. *Synonymovie* functions as an odd literary entertainment, giving life to new narrational possibilities in automated, technologically hybridized digital poetry.

Jody Zellen

Without a Trace (2008–9)[53]

Concentrating on urban culture, webdesigner and artist Jody Zellen's works often incorporate materials repurposed and processed from mass-media news outlets.[54] Zellen initially developed a remarkable manual technique that involves tracing over the newspaper and transforming both images and headlines/text into elementary but revealing line drawings.[55] She juxtaposed these line drawings in automatic static arrangements and in Flash animations. Eventually she began to use computer software to automate the tracing process. 'By selecting specific Photoshop filters,' she writes, 'I can make a news image look like a black and white line drawing. The computer's drawing uses an algorithm, whereas my drawings are based on subjectivity.'[56] Zellen's handmade drawings are imprecise; likewise the computer's drawings are also raw and abstract. Most present a discernible image but reflect the context, subject and activity unclearly; traced lines essentially portray delicate outlines of figures and words. Using the Internet as a 'sculptural space', Zellen embeds figurative copies of mechanized drawings and newspaper articles into several her works.[57]

Without a Trace, commissioned by turbulence.org, uses the *New York Times* as one of its sources.[58] Zellen randomly juxtaposes a live RSS text feed and an image taken from the daily paper, processed into a line drawing rendered by computer. She accompanies these elements with 'balloons' taken from 'Real Life Adventures' comic strip panels (stripped of text), a text excerpted from a comic, and a trace drawing of her own creation.[59] As seen below (Diag. 5.6),

Zellen aligns the balloon and trace drawings in a column on the left. Beside it, in a wider column, she inserts a larger machine-drawn image at the top, and sandwiches the newsfeed between it and a three line comic passage at the bottom. In 2009, the combination of elements changed once per day, resulting in what Zellen calls a 'calendar of juxtapositions – the days of which look similar but are not identical.'[60] While actively under construction on the WWW (2009), viewing *Without a Trace* involved no more than visiting the site and clicking a 'Daily Image enter' button.[61] Output drawn automatically appeared as a static projection on the screen, requiring no further physical input. Beyond accessing the interface, *Without a Trace* was non-interactive and contained no sound component. Today, viewers have three choices when exploring the work: to see a 'random version', 'daily image' or 'archives'.[62] Choosing 'daily image' when the piece was 'live' produced that day's static output, formed as described above and identical even after refreshing the screen. The 'random version' produces the same essential template, but when the browser refreshes, the appearance of every visual aspect changes except the headline.[63]

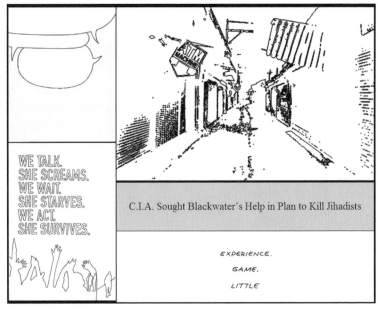

Diag. 5.6. Jody Zellen. *Without a Trace.*

As with other net-poetry, Zellen's *Without a Trace* contains variable language components. As seen in Diag. 5.6, text appears via the headline newsfeed, appearing in a grey box in the middle on right, and in the comic situated below it. Often, but not always, words appear in the drawings produced by Zellen. Automated drawings (top right) sometimes feature incidental text, but the emptied comic strip balloons (top left) never contain text, though certainly invite viewers to consider its absence as part of the projected dialogue and imaginatively fill it.

In this example, created on August 19, 2009, the headline for the day, 'C.I.A. Sought Blackwater's Help in Plan to Kill Jihadists', refers to a current event involving the 'War on Terror' in the Middle East. Zellen's program always takes headlines from front page news stories, so the focus of this textual component typically concerns major global or national economic or political issues. Archived headlines for the week 15–21 August 2009 – the sequence including Diag. 5.6 – typically reflect Zellen's current events context: 'Retailers See Back-to-School Sales Slowing'; 'Competing Ads on Health Care Plan Swamp the Airwaves'; 'More Troops Are Sought for Iraq's Restive North'; 'Frailty Lingers in Housing and Producer Price Reports'; 'Turnout Seen as Uneven in Afganistan as Polls Close'; and 'Fed Chairman Says American Economy Is Poised to Grow'.[64] No matter what method of comprehension a viewer practises, *Without a Trace* addresses significant political and/or cultural issues due to this particular authorial design decision. Since accompanying texts tend towards the graphical, Zellen creates a circumstance in which the viewer's receptive tasks rely on interpreting visually connections brought forth by the contrasting arrangements. Incidentally, as in Tisselli's *Dadanewsfeed*, or looking further back, to the written poetry of Allen Ginsberg circa *The Fall of America*, the litany of headlines reads like a poem guided by significant news of the day.[65] If we read the lines above in a linear fashion, the first line sets the stage for a series of observations, mostly involving various types of unrest, which here happen to lead to a positive determination at the end of the week (i.e. 'Economy is Poised to Grow').[66]

The inclusion of three short lines of text in hand-written font beneath the headline help make Zellen's work more poetic.[67] Text beneath the headline in Diag. 5.6, excerpted from a comic,

'Experience./Game./Little', always appears in three cryptic lines. If we compare this text to the headline text, it seems to suggest that the experience of a game or experiencing a game – any sort of game – is something small in comparison to the sources of turmoil found in the headlines or in Zellen's drawing. In this manner *Without a Trace*, as a poem, juxtaposes multiple, inherently separate, text areas. Viewers must work, against any apparent discrepancies, to make unifying connections, and then attempt to integrate other available information (textual or graphical) into a conceptual model or understanding. In the week sequence referenced above (August 2009), the following sequences of words appear: 'government/ change/officials'; 'sadly./choose/left'; 'depends./made/face'; 'speak/ this/talking'; 'experience./game./ little'; 'this/kill/finish'; and 'will/ sentences./sea'.[68] These snippets from comics, in spite of quirky punctuation, can be seen – especially cumulatively – as poems that contain messages readers may apply to the surrounding combination of images and perhaps connect thematically to headlines.

Zellen's drawing in Diag. 5.6 includes six lines of text (more than usual) clearly appropriated from a human needs campaign; illustrating cause and effect relations, it encourages people with resources to help those who do not. The illustration also contains a solitary figure walking away (on the right) and raised hands. Reconciling the human needs text with the 'game' aspect presents little difficulty. A clear concern for *polis* arises in this circumstance: introducing the domain of human suffering puts any sort of game, or gaming, into its proper context: in the category of folly or non-essential activity. Beyond these characteristics, this example shows three different-shaped comic balloons and an automated drawing from the *New York Times*, which appears to be a desolate alley that coincidentally captures clearly a sign for an ATM Machine.[69] Text surrounding the headlines, when and where it exists, does not always immediately connect to cultural or political concerns, so it becomes the viewer's challenge and task to fuse textual elements. Chance appearance of the ATM sign only serves to solidify the role that capital plays across all walks of life, thereby providing a certain slant for a unified reading of the materials. Zellen's empty comic balloons, as always, represent visually space that could be filled in with comments about the combination of these variables, a dialogue about the issues at hand to be completed by the viewer, or perhaps indicate that responses to the conditions conveyed by the textual events cannot be found.

As indicated, *Without a Trace* features text beyond headlines. These other 'voices' and ghost voices and images – part of every sampling – become a type of virtual oracle or medium that uses purposely disconnected attributes as a response to fragmentary conditions experienced by denizens of contemporary culture. If only fractionally, Zellen's project traces random aspects of global culture – as filtered through news and other media – that can be re-traced in the archive configured by the site. By exploring the archive, the reader can track not only a course of (then) current events but also gauge successes and failures of cultural initiatives and predicaments, such as the war on terror or the economy, as well as consider and interpret what Zellen's 'oracle' had to say on any given day.

Of her work *All the News Thats Fit to Print* – which exclusively uses image and headline feeds from the *New York Times* – Zellen writes, 'The new juxtapositions become wrong, sad, funny, inexplicable, and often to the point.'[70] Its unique graphical qualities aside, the fact that resources inscribed in *Without a Trace* include and transcend the newspaper while not necessarily broadening its focus allows for expressive variation and capacity within the 'wrong, sad, funny'. Information taken (and reduced) from the comic arts particularly promote the artistic qualities of her work, as more influences and considerations for participation spice up the equation. Minimalist, slow and abstractly pictographic, *Without a Trace* maintains focus by connecting to a news source without relying on that source entirely to incite a viewer's reaction or convey meaning. Although viewers could generate multiple different samples by using the 'random version' section of the piece, only one sample of output was produced per day. Viewers could have difficulty bridging incongruent materials projected by *Without a Trace*, but its static presentation eliminates the type of chaotic shifts that occur in many other net-poetry works, and this characteristic provides ample time for contemplation.

The notion of 'trace' or 'tracing' plays an important role in Zellen's work: she traces over images to make new images and traces across the (transitory) news headlines to acquire transmitted content. In her project, Zellen mixes together fleeting information – temporarily highlighted by the news media – and reorganizes it into a meditative poem that gradually progresses over a year. *Without a Trace* changes daily, yet contains no kinetic properties; its transformations occur behind the scenes in the computer's processors and in the solitary acts of the poet. In her comments on the piece, Zellen claims its construction, 'takes as its point of departure the idea of a daily ritual.'[71] Thus, despite the computer's intervention on several levels, a human aspect and scheme remains at the core of the work. As news flows across the network during a given day, headlines and news images always shift. By uniting initially standalone text and creating new juxtapositions while at the same time introducing new variables, Zellen's combinations broaden and deliver a type of interconnected awareness that singular representations cannot.

Now let us step back and take another look at the case studies in this chapter as a whole. As we have seen, Lehto, Tisselli and Andrews impressively base their digital poems on perpetually shifting core foundations, thereby producing results unobtainable by print authors. Queneau's famous slotted poem is expansive, but its materials are fixed. Using the network's processes to attain varying output of expression extends collage technique at its roots. Kinetic manifestations of pastiche language and image, *Google Poem Generator*, *dbCinema*, *Synonymovie* and *Without a Trace* are historically unmatched due to their utterly contemporary methods of acquisition and projection; treatments enabled by computer and network bring new possibilities to literary endeavours. Lehto's work, using the world's largest information index, widely stretches the boundaries of information constraint within a poem; never before has such a broadly informed resource that draws from all aspects of human culture been so readily available for an author to use. The same holds true for Tisselli's work and *dbCinema*. No doubt many poets have made cut-up poems with the nation's 'paper of record', but how many of these poems have been seamlessly fused with a range of other materials? Only digital technology makes it possible to present a daily serial poem that features a combination of hand-made and automatically generated components. By incorporating network mechanisms, we discover unforeseen connections between words, images and all aspects of culture – how people, places and ideas can be presented and associated through language and technology. As chronicled above, with these inventions and ever-changing works, we uncover limitation, surprise and poetic thought.

The net-poetry discussed in this chapter demonstrates how the WWW plays an intrinsic role in the creation of new forms of poetry and literary artwork. An expanded study of creative and conceptual pieces of digital poetry indigenous to the WWW could also include an investigation and discussion of works made by some of the pioneering artists in the genre, including Philippe Bootz and Cayley. Before we conclude this chapter, let's look quickly at Bootz's "Désir Insuffisant"[72] and Cayley's *Imposition*.[73]

Bootz's "Désir Insuffisant" ('Insufficient Desire') is a WWW-conceived but 'unpublished' work that does not contain any material at all. It cannot be seen, even on the WWW, but contains processes that cannot exist out of the WWW. Bootz prepares only a paratext entitled "à propos de désir insuffisant" ('About Insufficient Desire'), in which he demonstrates the work's properties. Describing the genesis and concepts of "Désir Insuffisant," Bootz writes:

> when it was created, I had no works on the Web (except "Haiku puncture"). Some people … wanted to search [for] some of my works by using Google. Naturally they find nothing except how to obtain works from the site of MOTS-VOIR, which always appears first when searching on 'Philippe Bootz.' This way is very simple: it is sufficient to contact me or the publisher 'in the real world'. But these people did not make this [connection] and continued to search. So, their desire to read my works was real and this desire was created by the existence of the works, but their desire was insufficient to switch in the real world. So their navigation between the documents given by Google was for me material proof of interactivity with my work that is really web-based. The work itself is the absence that generated this navigation. Without this absence, there would be no navigation and this navigation is relative to the meaning of this absence (this meaning, for me, is the indication of the real digital nature of my works the people made by their navigation and the real autonomy of the works that are not 'closed' to the digital world). I named this absence 'Désir Insuffisant' and I signed their navigation as my work![74]

In the paratext, Bootz describes how he uses the classical rhetoric figure of inversion: he inverts media and process and content of node and navigation. In effect, "Désir Insuffisant" is a WWW-work but not a digital work: a poem, without words, because it asks for a reading of, and response to, real information. Readers who did not "trust" that Google's search engine was giving them factual information continued to search for Bootz's work. Its absence on the WWW was read as a fiction rather than an accurate document. These fruitless searches make up the contents of Bootz's poetic construction.

Imposition, writes Cayley, is a 'networked performance of an evolving collaborative work engaged with ambient, time-based poetics and harmonically organized, language-driven sound.'[75] The work's audio and visual display emerges as a result of networked-linked computation and textual manipulation. During a performance, twelve different programmed QuickTime movies ('listening movies') prepared by the author become available for download and playback on computers operated by participants.[76] *Imposition* incorporates texts and procedures used in *translation* (Cayley, 2004), and thus continues his 'investigation of iterative, procedural removal from one text or language to another'.[77] As a performance transpires, *Imposition* presents passages from source texts in one of three shifting states: surfacing, floating or sinking. Moreover, passages appear in one of three changing language states: German, French or English. As a passage dissolves in one language it may appear in another; distributed across the WWW, this creates a 'transliteral' musical performance.[78] Cayley's description of the work details how a particular local rendition, consisting of an arbitrary number of listening multimedia devices, results from events happening elsewhere on the network:

> *Imposition*'s main display shows the four transliterating passages on a large projection and broadcasts their states.... While the main movie is running, listener movies will track one of the four passages, but will do so in a single language (as selected at download time by the participant who plays the 'listener').... These movies also play looping musical samples of human vocalizations which harmonize both with the main display and with other linked and listening movies. The selection and triggering of samples also reflect their linked passage's buoyancy.[79]

Although it doesn't stream from the WWW per se, the audiovisual content generated on an audience member's terminal in real time depends on textual and other events that occur on a server located elsewhere on the Internet.[80] Described by Hayles as a 'multimodal collaborative narrative', *Imposition* represents a distinctive form of technological transmission for digital poetry.[81] When networked participants bring together the elements engineered by Cayley during a performance, they conduct a verbal and visual ambient concert of linked instruments; the individual movies respond to a single complex process and yet, interestingly, the devices will perform somewhat out-of-sync since the downloaded movies play at difference rates and are not started simultaneously. Nonetheless, due to the compositional harmonic strategies of Cayley's collaborator, Giles Perring, the samples harmonize in real time without need for the kind of orchestration normally found in symphonic arts. Presentations of *Imposition* are one-of-a-kind occurrences shaped by the number of participants and how applications linked to the network respond to their virtual networked stimuli when activated by viewers.

Offering a partial conclusion to all of these materials, I relay a comment Ted Nelson makes in his book, *Geeks Bearing Gifts*: 'Google now owns all of literature'.[82] Whether or not this is true – and presumably Nelson primarily refers to a lawsuit giving Google the right to print pages of any book – the fact is that Google and other network repositories and resources now play a major role not only in the distribution of literature but also in the *making* of literature and art. Obviously not all WWW-reliant pieces of artwork rely on Google or search technology, but several of the programs discussed above seek and gather information via Google in order to craft automated and mental processes into pieces of verbal-visual artworks.

As we have seen, poetry readers and writers embrace the contents of and get energy from the WWW. The raw material of the network, combined with authorial invention, engineers a creative, potentially transformative experience within the materials provided. By disrupting a typical flow of information, these digital poems, like others, bind random and strategic possibilities for expression. Language-based Google poems may be less dynamic in terms of content, but they introduce surprising illogical elements to be consumed and finessed by readers and/or authors as crafted expression.

Andrews' and Tisselli's work compile unique visual properties and layers of imagery atop one another. Digital poetry has become more cinematic, a fact reflected in metaphors used in titles by Tisselli and Andrews and reflected in works across the field, including *Slippingglimpse*. By usurping a commercial apparatus, a network containing millions of texts and images, the above artists access and then synthesize openly available materials into the body of costless digital poems.

6

In stages, on stages: attentions in digital poetry

In the Preface to *Unoriginal Genius: Poetry by Other Means in the New Century*, Marjorie Perloff, eminent scholar of contemporary poetry, makes a much-exaggerated observation. 'There was, it is true, much talk of the possibilities of "E-poetry" – poetry written and formatted for the new electronic screen,' she writes, 'but E-poetry never quite got off the ground, the compositional process of an E-poem (however much animation might be used) not being essentially different from that of a "normal" print poem.'[1] Her commentary, understandable from a perspective that views all electronic works as collage or pastiche, reflects an off-base position for several reasons.

Firstly, subjects of study in this book demonstrate how compositional processes of many electronic or digital poems diverge from and are dissimilar to those found anywhere in print. As I have discussed, a digital poem's reader often determines the content of a work – a kind of 'reading' virtually impossible in print poetry – by modifying virtual structures provided by the digital poem's author.

Mutability in the appearance and language transmitted in digital poems locates them in an altogether different modality than works based in print media. How can a page-based reader consume or absorb a poem by destroying and transforming words, as do viewers of, for example, Andrews' *Arteroids*? Procedures of text acquisition, imagination and invention may be equal for all human writers, although engineering seamless, sometimes controllable non-linearity in screen-based works, while Dadaistic on the surface, at very least requires technical knowledge, skills and processes more complicated than those used to devise the arrangement of words on a page. Contemporary digital processes, associated with the machinery that delivers them, play a role in all aspects of the presentation of content – the poetic (textual) and material (electronic appearance).

Furthermore, to state that digital poetry's greatest advancements since the WWW's onset have been in dissemination,[2] rather than in form, is to tell a partial truth. My case studies have iterated and reported on progressive dynamics emanating from complex forms of poetic expression embodied in these works in ways that contradict Perloff's viewpoint. To suggest that digital poetry as a discipline has come and gone, thus dismissing it for lack of sophistication, unfairly considers a fledgling genre of artistic expression still in the process of getting off the ground. To be sure, Perloff and others who dismiss digital poetry are more likely underplaying its vibrancy and complexity due to a lack of indoctrination or participation – why else reject it as something that barely happened, when so much evidence to the contrary exists? Unquestionably, digital poetry *is* happening. Digital writers have, for more than five decades, constantly found lively ways to present poetic expression through computers by employing methods inaccessible to print-based writers. The use of constraint and visualization in innovative processed writing certainly occurred before the digital age, but these procedures and techniques, in that prehistoric age, so to speak, only partially correspond to the groundwork laid out and employed by digital poets thus far.

Resistance to investigating more fully digital poetry's ramifications and merits perhaps results from the fact that digital poems intensely challenge the comfort and confidence of a readership used to the page where a poetic document sits still and can be fully absorbed. With electronic works, however, such luxuries rarely exist. The most avant-garde print works may puzzle and even disorient readers, but the materials print-based authors present are always fully available.

Yet in certain works discussed above, such as Angela Ferraiolo's *Map of a Future War*, encountering all the verbal structures and other components engineered by the author simply defies practicality. Indeed, WWW-reliant works utilize perpetually shifting materials, and may never project the same content twice.

Digital poetry's compositional processes may be similar or even identical to those found in print-based works. However, the application or rendering of the processes makes an inherently distinctive impression on its viewers, inducing varied and atypical effects. Perloff, though, overlooks such operations. By only acknowledging the presence of animation, she fails to account for many other noteworthy qualities of digital poetry. Her brief commentary on the subject neglects to notice, address or analyse many elements that make the genre a truly distinctive pursuit.

Typical complaints about works appearing on the WWW involve disconnection – broken links, missing information, software incompatibility or corruption, and the inconvenience of (and insecurities involved with) downloading or upgrading applications. Most users of the network adapt to such nuisances, and accept the network's sometimes inconsistent functionality. Technical grievances by insiders such as Ted Nelson, on the other hand, suggest the WWW itself presents inadequate conditions for delivery of artworks. Nelson criticizes the 'Web's iron browser standard' which in his view imposes undesirable limitations on documents.[3] 'The Web', writes Nelson, 'is about fonts, hierarchy and outward links only'. Here we see a view that also presents an unbalanced, absolutist perspective not wholly supported by the case studies offered in previous chapters.[4] Nelson desires interfaces that inventively extend to new areas, outside the boxy browser, and believes delivery of content could be streamlined and made multiplex, in the sense of many signals being combined into a single transmission circuit or channel. He is not so far off here. Contemporary developments, such as augmented reality (see below) and net-poetry initiatives (Ch. 5) already begin to explore these areas of creative inquiry within the context of the WWW.

To Nelson, the display apparatus also plays a role in restricting electronic text. In *Geeks Bearing Gifts*, he explains we rely on a hardware scheme designed for convenience at Xerox-PARC in the 1970s that imposes limits on expressive potential. Nelson's apprehension stems from a belief that graphical user interfaces 'took away the right to program', changed the appearance of everything, and altered the expectations of readers.[5] This development led to the popularity of personal computers, as it made them easier to use, and profits for corporations, but worked in opposition to the command-line interface, locking users into a 'a paper-like rectangle'.[6] The WYSIWYG (what you see is what you get) interface runs the risk of relegating the computer to being 'a paper simulator' rather than an expansive yet unified, commercially inventive broadcast ecology whose visualizations could be individually customized.[7] Have the circumstances on which Nelson bases his observations impeded digital poets? To the contrary, artists preparing works using tools now available certainly reach previously unattained textual achievements.

Nelson's contentions contain flaws, particularly because works made with tools such as Java require programming, and manually creating and editing code plays a significant role in JavaScript, Processing, Flash's 'ActionScript' (which has become increasingly automated) and other methods of composition. Regardless, as a result of the predicament outlined by Nelson, electronic literature at this juncture has grown out of aesthetics formed by print, even if its best titles modify, transform and sometimes ignore tenets of historical paradigms. Complex and unconventional efforts by digital poets constructively and fruitfully respond to an unstable, materially limited state of existence, and discover inventive alternative uses for commercial software and hardware purportedly seeking to contain them.

Beyond problems caused by imperfection, browsers, hardware and other issues, a larger fact looms. 'Our future in pixels', writes Rasula, 'will not be monumental.'[8] Material conditions for digital poetry greatly diverge from the model proposed by the 'Grecian urn', as something beautiful and lasting. Rather than build something heavy, set in stone, electronic poets rely on a technological equation (electricity, hardware, software and more) that must perform concertedly or does not function (i.e. cannot be seen) at all. In online works, where the dysfunction of any number of digital components can disable a work, this is especially true. Whether an 'outage' is temporary or permanent, unreliability is a serious characteristic of digital poetry; disappearance is a certainty. During the course of this book's composition, multiple works I studied became inoperable on the WWW.[9] Flash is known to have longevity issues. For instance, as Larsen reports, new Flash shockwave (.swf) players 'may run into weird code with the older flash .swfs' and cease to play them.[10] Artworks can be updated or rewritten to new specifications, but are not really the same piece. While works may be viewed as 'well-wrought', they are not solid, in no way everlasting, and may be available only temporarily. '*Ars longa, vita brevis*' simply does not apply to digital works, especially on the WWW. Future obsolescence of the technologies involved, and even the WWW itself, is a given, as implied by the title of a September 2010 article in *Wired* magazine, 'The Web Is Dead. Long Live the Internet'.[11] Even the 'Wayback' machine – already incomplete – will eventually fail and a huge amount of creative and other efforts will disappear.[12] Most digital artists hold different aspirations from those of the ancient Greeks, Romantics or Modernists; permanence is not among them. With digital transference of expression, and developments such as interactivity, the artistic foundations for poetry change.

Main areas of inquiry in the field of digital poetry at the time of the 2009 E-Poetry Festival focused 'on practice and reflection, close reading and code' (Borràs); Bootz noted also the attention to the literary nature of digital poetry, asserting his view that teaching and translation were challenges he hoped to see more engagement with, a view endorsed by Cayley, who observed that translation implies literary practice and shows the importance of language in the genre. Papers by Bootz and Bouchardon addressed the problem of obsolescence by emphasizing the importance of archiving. Bouchardon's 'Conservation of electronic literature works' discussed 'media decomposition' and initiatives prioritizing the preservation of digital writing.[13] For Bouchardon, technology presents the possibility of heritage preservation, but in the current predicament no definite solutions exist for problems caused by new (and old) formats, proliferation and heterogeneity in form. Bootz introduced the ARCHIPOENUM project, designed to index documents related to digital work using different ontologies and procedural models.[14] Both papers indicated establishing archives as a priority in the field of digital literature.[15] The WWW is a profound – but temporary – zone of creative exchange. On- and offline contemporary artists experiment in countless interesting ways, but something seen (or experienced) one day or one year may be inaccessible for any number of reasons the next. This instability discomforts several serious practitioners within the community, as reflected in the archival/conservation priorities enumerated by Bouchardon, and by digital recovery and retention efforts made by others.[16] At present, Flash works cannot be viewed on mobile devices such as iPhone – an issue of corporate territorialism – but to imagine Flash animations, or even the entire WWW, becoming extinct is not beyond comprehension.

In spite of Nelson's reservations and questions raised about the durability or longevity of forms and disconnection from the literary, tools at the disposal of digital authors succeed – and should be studied – precisely because they were, as reported in *Geeks Bearing Gifts*, 'chosen for you because they were simple'.[17] Digital poetry transpires when multifaceted operations craft verbally oriented expression on machines built for scientific and commercial tasks. Nelson's dissatisfaction with the rigidity of the WWW relates to the study of digital poetry because, as he writes, 'Technical documents may be traditionally hierarchical, but … poems are not.'[18] Since many of the titles explored in this book manage to resist reflecting this trait (i.e. they lack hierarchical appearance), digital poetry triumphs because authors have found ways to disrupt the conventions, if not the identity of the medium, by developing captivating creative methods in spite of obstacles. Electronic literature transforms the 'page', as Hayles writes, into 'a complex topology that rapidly transforms from a stable surface into a 'playable' space.'[19] These digital developments do not reflect an artistic digression, and incite no cause for alarm, but rather signal achievement in creativity – however temporary such successes may be.

On an aesthetic register, Nelson offers a significant grievance about the WWW: 'NO BRANCHING MOVIES'.[20] Although theorized by Nelson as early as 1974 (see *Dream Machines*), little research has focused on this modality; this form of hypermedia eludes the attention of most filmmakers, authors and multimedia artists. Few of the millions of videos now available on the WWW through YouTube, Vimeo and other resources practise 'branching' from within active narrative activity; viewers proceed from one video to the next only upon conclusion. Hollywood and mass culture seem uninterested in anything beyond traditional linear movies. Yet, as illustrated in the case studies above, Ferraiolo's *The End of Capitalism*, produced after Nelson's book was published, represents a significant literary interest and turn toward this non-linear artistic possibility. Whether desired by the public or not, electronic literature proceeds to re-imagine the manner of textual delivery.

Despite its 'grotesque programmer's interface,' writes Nelson, 'Adobe Flash rules in interaction today.'[21] As made clear in the investigations above, Flash can be used as an interactive textual synthesizer permitting the incorporation of all forms of media, and enables internal and external links, in WWW-based works. Presumably, in time, more types of branching video-based productions envisioned by Nelson (explored already by artists such as Ferraiolo and Manovich), built with either Flash and/or another program, will emerge within and outside the boundaries of the literary. In retrospect, the era studied in this book might in fact be seen as the 'Flash period' – although as works explored above demonstrate, many other programmatic options exist.

An artist creating with Flash prepares materials on a stage, using a timeline. Digital poetry as a WWW-based, networked form happens *in* stages, *on* stages. Few of the refined works discussed above require singular responses from viewers/readers or user/players, who focus attention on the screen where authors have orchestrated materials. Both the composition of works and viewing them involves multiple steps and visual scenarios, and, as Memmott crucially observes, 'it is essential to understand each digital poetry application as an environment or poetic microculture with its own grammar and customs.'[22] As the history of digital poetry clearly evolves in stages, viewing and receiving content within unique individual works also occurs in different phases, as we learn to blend verbal, visual and sonic components to be absorbed as a whole. 'If the expectation of a reader-user is that she will discover the secret of a particular piece by abstracting its elements,' writes Memmott, 'isolating the verbal from the visual, the environmental grammatology of the work is lost and the outcome is not a close reading but a partial or close(d) reading that depletes the work of its poetics.'[23] Orders of presentation can be variable or multiple, and progress through the works transpires in successive, programmed, integrated phases.

Poets involved with programming occupy an exciting position in the arts. Since computers can be used as language machines, and visual and sonic computing closely associate with language computing through software, all of these methods, especially when digitally fused, involve inscription and strengthen the possibility of hybridized, performative writing, giving the creative writer who combines other artforms (as well as information technologies) potent advantages in staging texts.

Modalities of practice in digital poetry have expanded in the past decade. Expressive inclinations diversified, becoming completely re-directed for some artists; pre-WWW aesthetics advance with refinement, and new artistic interests can be charted. Several significant 'prehistoric' digital poets made computers central to their craft, producing creative work online, where they find unique directions in which to cultivate expression and become leaders in the global community of digital poets. The expressive whereabouts of these works reflect the span of possibilities available to anyone involved with creative programming; activities highlighted above – by a selection of artists who practise various forms of digital poetry in English – demonstrate numerous possible directions and reveal many important details about the historical trajectory of digital poetry.[24] Dominant modes of presentation, projected and participatory, are clear. Attributes common to works are detectable, but each example makes different cognitive demands on a viewer's attention by delivering artistic stimulation in unique and varying ways. Absorbing content entails studying and learning to use inconsistent interfaces. Whether materials rapidly appear and disappear or whether they linger on the screen, learning to navigate demands concentration – something readers must grow accustomed to expect in any electronic literary work. Finite holistic strategies for experiencing and engaging with digital poems otherwise resist assertion: built-in conditions of incompletion characterize many titles, and the pace of reading sometimes prohibits full absorption and processing of materials. The rapid pace of *Mesostics for Dick Higgins* intends by design to demand acquisition of information beyond conventional comprehension. Other titles exclude the possibility of complete readings (except in a conceptual sense) due to excessive scale (*Internet Text*), duration (*wotclock*), idiom of presentation (*Arteroids, Map of a Future*

War), and size of database (*dbCinema*). Carolyn Guertin, considering this predicament, writes in celebration of it: 'the remix factor of hypertext and newer kinds of digital literary forms over-determine the organic structure, for the author has to assume that we will not visit the whole of the text in our travels, and so has to prepare for every contingency – as in life.'[25] Viewers thus need to adjust cognitive attention to appropriate frequencies, which differ between pieces, and try to read the work – and 'read between the lines' of the work – for theoretical/abstract indicators in addition to gleaning content via direct contact with it. If a work maintains our interest, we find ways to discern its purposes and discover its functionality on a more surgical level. Various factors, including multiple layers of information obstructing text(s) from view, prevent complete readings – as in Baldwin and Sondheim's Second Life constructions. Cloaked text does not always emerge. In most of these works, only fragments of the overall content can be seen at once. *Carving in Possibilities*, *Stir Fry Texts*, and Rosenberg's Diagrams intentionally exclude content from view; *wotclock*'s temporal constraints more extremely limit access to its overall content in the moment. Other titles do not obscure materials, but viewers cannot always control what they see, e.g. in certain examples, content cannot be revisited without starting from the beginning.

Participatory or 'played' works (e.g. *Arteroids*, *Carving in Possibilities*, etc.) give significant control to the viewer, where transformation or modification of given materials leads to the artistic result. Beyond mental and other sensory factors, physical movement of the hand plays an important – if not always strenuous – role. Even if aspects of Bouchardon's works indicate new approaches to input, manipulating materials presented primarily involves the mouse and keyboard. Digital poems present themselves as pieces of a puzzle to be solved; viewers make selections and put pieces together while engaging with forms and content presented through its interface and media. These works feature verbal elements, text delivered as writing to be read, but frequently include media attributes – a fact that puts distance between contemporary and historical generated and visual works.

The same circumstances apply for each digital poet working on the WWW, and we can (and should) appreciate that authors construct poetic materials in so many different ways, even when using the same software programs. Production of digital objects made 'by hand' – when authors manually produce entirely unique and sophisticated textures instead of using pre-packaged, smooth, automatically rendered or photographic objects – sets some of these works apart (e.g. *Seedsigns for Philadelpho* and Second Life environments). Using commonly found fonts, plain text, photographic images and conventional controls such as hyperlinks to carry digital expression capably and artfully performs expressive tasks, but titles containing what might be called 'folk' (handspun) elements very effectively attract a viewer's attention due to surprise at seeing something unexpected and original.

As might be expected, some authors greatly expand the scope and mechanisms of their work on the WWW, with extraordinary refinements and expansion of artistry. Others remain relatively close to already pioneered modes of expression. For the past two decades, aND has actively practised multimedia artistry, and his work now takes advantage of the WWW's fluid transport of animated/videographic material. Examples of works by aND discussed above – although not entirely indicative of his recent efforts – contrast with those he produced before the WWW due to their cinematographic and non-participatory approach.[26] Using sophisticated animating techniques enabled on the WWW, aND moves away from some of his original inclinations and extends his unconventional aesthetics into new areas permitted by the technology now available. In *Seedsigns for Philadelpho* he selects and animates anagrams; in *Mesostics for Dick Higgins* he animates mesostic-form poems, which also involve selection and choreography of poetry. Andrews has completely left his static past behind in favour of producing dynamic, interactive poems. Each of Andrews' titles transcends the static models of his early practice into interactive realms. His work, as before, includes significant visual properties, which extend into new areas especially through *Arteroids'* incorporation of gaming motif. *Stir Fry Texts*, *Arteroids* and *dbCinema* are participatory; viewers make unique instances and sequences of poetry appear on the screen and provoke its kinetic properties. Cayley, in his processes of 'ambient poetics', produces interactive, animated works (textual and graphical), involving techniques he has described as 'transliteral morphs', but in the mainly projected *wotclock* he repurposes – and makes more elegant on the WWW – a previously engineered work, to which he adds videographic elements through QuickTime. He combines his interest in cultivating a clock that issues poetic utterances with contemporary tools enabling the packaging of his mechanism with spectacular, interactive photographic imagery.[27] While conceivable that *The Speaking Clock* could have been made more vivid at its moment of construction using a program such as Director, someone with the required skills can now use commonly available and affordable programs to make high-quality cinematic works. Larsen continues as a hypertext artist but presents a very different type of hypertext in *Carving in Possibilities*. She forsakes the traditional model of disparate lexia (verbal fragments), and essentially makes all of her work's verbal information – combined with sound effects – directly accessible on a common backdrop. Further, she simplifies the viewer's tasks by removing the need to click to proceed to new textual information. Larsen completely does away with her pre-WWW approach to hypertext, whereas Rosenberg's Diagrams (also hypertexts) follow a steady historical path, the main difference being the program used as an authoring tool. Of all the works presented in chapter 3, the surfaces of Rosenberg's poems have changed the least. Some aesthetic modifications do exist in these recent examples of his Diagrams, but they closely resemble the type of works he created before the WWW. Sondheim has been nothing if not a versatile artist, who has always produced works in multiple forms. In some respects he continues to cultivate the experimental modes he has worked on in the past in generated, interactive graphical works, but also advances the qualities of his expressive interests – especially in his collaborations with Baldwin and Strasser – sometimes due to technical advances enabled by Flash and the visceral 3D sensibilities of Second Life.

Efforts of contemporary practitioners perpetually redraw and extend the aesthetic maps previously established for digital poetry. Among other advancements – not the least of which has been time for consideration of the subject – tools used today differ greatly from those available during the first three decades of poetic practice conducted via computer operations. While artists could (technically) and did (occasionally) use Director, QuickTime and video in digital poems produced before the WWW, most of the techniques employed for poetic production in works discussed above (html, Flash, dhtml, Cascading Style Sheets, JavaScript, Squeak and Second Life) emerged since the dawning of the network. Certain technologies used on the WWW, such as CSS, give artists tools to engineer high levels of uniformity and functionality within individual works, but the range of techniques now being used more importantly serve to propel flexible synthesis and variation – overall diversification – within the genre in a way previously not possible. The earliest iterations of digital poetry were confined and unified by their use of programming code and ASCII text; everything more or less looked the same until graphical user interfaces in the 1980s began to vary the presentation of kinetic and visual works that emerged in the late '60s. Artists working in the field still emphasize permutation and mutation in appearance and content, and use databases containing language and other materials, but now have a wider range of opportunities to devise poems by altering the appearance of texts as a whole or by making adjustments to words at the level of the letter or interface. More dynamic, spectacular alterations, amendments and supplements to texts increase the interplay of artistic variables. Artists previously practised the types of textual mechanics now enabled through procedures such as html and JavaScript, but operated at a disadvantage due to the newness and unfamiliarity of cultivating hybrid expression, and also by limitations posed by the inaccessibility to the stage on which they presented materials. Operations digital poets now conduct with Flash were previously possible on some registers, but the labour and cost involved with creating dynamic alphanumeric and graphical animations prior to the 1990s was prohibitive and would not have garnered extensive audiences unless they were televised. In the WWW era, these impediments disappear.

Artists are not bound to one method of approach: they may impart interactivity in one piece, i.e. make participatory works, and project content onto viewers in another. Works discussed above, in sum, present the gamut of historical approaches; each of the prehistoric forms appears – although the automatic spontaneous generation of poetry is less privileged due to the vast array of generative possibilities now available. Text-generators played an enormous role in digital poetry before the WWW and automated composition of textual result remains central to the crafting of many digital poems on the WWW, usually as one of many components. Works by Carpenter, Cayley and Sondheim generate output from complex databases, and generative properties emerge in the narrative path for the viewer in *The End of Capitalism*, in Strickland's text-image association, and certainly all works highlighted in chapter 5 except Sérandour's. The role of generation has, however, shifted towards media computation, which often inserts pre-processed language. Works explored in the case studies discussed largely explore animated conditions for text, and non-animated works prioritize interaction.

Chapter 3 introduces works made by digital poets active in the field for at least two decades. Leaders by example, these artists have created works exemplifying qualities that will become standard models, if not already occupying such a status. Digital poetry has become identified as a genre in which works are prepared exclusively for mediated presentation: works that can only properly function via computer screen. Works discussed in the case studies, with the exception of portions of Sondheim's *Internet Text*, depend on compositional software and computer hardware.[28] Words in aND's *Mesostics for Dick Higgins* and in other titles originate in books or other print-based sources, but are then manipulated by some aspect of technology employed by the author as in Rosenberg's *Diagram Poems*. Even if 'written' words populate their core, the final product – the end result and sum of expression – could not exist without computer mediation. Poetry has always been an intertextual form, and now with computers becomes an intermediated one.

Works produced by artists in the WWW era clearly attest to the evolution of digital poetry during the network's first decade. Many of these contemporary productions embody the genre's 'prehistoric' tendencies, while others overtly reflect refinements within forms outlined in my previous study: Carpenter's *Erika* and Cayley's *wotclock* (in part) generate text, Mez's work projects (kinetic) animation, Flanagan makes 3D visual poems, Ferraiolo's *Map of a Future War* and Strickland and Jaramillo's *slippingglimpse* represent refined contemporary hypertexts, and so on. Increased levels of programming as well as advancements in technology bring variant conceptions. Bouchardon's sensory inputs in *Touch* and *Loss of Grasp* represent altogether new types of interactivity. In Ferraiolo's

The End of Capitalism a type of hypervideo has come to fruition. As predicted in *Prehistoric Digital Poetry*, digital poets have appropriated the mechanics of gaming. The act of reading changes due to the new interfaces, as Hayles (paraphrasing Markku Eskelinen) points out: 'with games the user interprets in order to configure, whereas in works whose primary interest is narrative, the user configures in order to interpret.'[29] An authorial drive to create games, beginning with *Arteroids*, is certified by digital poem-games such as Nelson's *I made this. you play this. we are enemies*. Even more significantly, in terms of compositional advancement, use of online resources, such as WWW-based image and/or text searches and RSS feeds, emerges in works prepared by Andrews, Tisselli and Zellen.

Digital poems exhibit unique properties in a now persistent (but once fresh) platform for the genre. Many recent works embody traits of digital poems composed prior to the WWW, but works that robustly utilize the unique properties of the WWW – particularly those discussed in chapter 5 – significantly differ from earlier efforts because of newer technologies such as Google's image search features and other communications protocols (i.e. APIs) and techniques (e.g. 'screen scraping'). Hayles writes of Strickland's conversion of *V: Wave Son.nets/Losing l'Una* into *Vniverse*: 'when a work is reconceived to take advantage of the behavioural, visual and/ or sonic capabilities of the web, the result is not just a web 'version' but an entirely different artistic production.'[30] A parallel observation can be made of a work that cannot exist without the WWW: the results of these experiments do not simply activate WWW versions of poetry but an entirely distinctive type of composition emerges. The case studies focus on contemporary practices and new directions in digital poetry, showing how hybridized, interactive text-image-sound forms typically encompass more than a singular aspect of historical forms and typological qualities. Some reflect innovation, some borrow methods from the past – but do so in increasingly complex ways that should not only be acknowledged and recognized but encouraged and rewarded. By now, digital poets have unquestionably discovered a range of valuable creative fabrics suitable for use among contemporary media possibilities. Such artistry has been successful, whether or not it has approached its peak yet. In recent decades a monumental transformation in textuality has occurred, and although works have become more complicated, we are past the turmoil-of-existence-and-aesthetics stage and realize the range of degrees to which writing has expanded to include all kinds of fused forms. Further, artists will continue to cultivate additional aesthetic areas, involving more physicality – works attuned to bio-feedback and different sorts of corporeal interaction – on- and offline.

As observed previously, most WWW-based work, built on the foundations of digital literature, does not require a network connection. Works utilizing special properties of the WWW inevitably use a connection to the Internet. How digital poetry – as represented in the case studies – uses the abilities of the WWW to propel expression is thus an important matter. None of the titles published in *Electronic Literature Collection*, Vol. 1 and few of those appearing in Vol. 2 – the authoritative resources representing the genre – fundamentally require the WWW.[31] Poems inseparable from the WWW represent the latest – but by no means final – chapter in the genre's history, and these works will disappear most quickly as network/delivery systems continue to evolve.

Looking back, certain disciplinary interests – types, methods, orientations – persist. For example, major themes articulated by aND following E-poetry 2003 – 'e-vispo, typography, text, programming, codework, soundpoetry' – continue to be integral to artistic engineering practised by digital poets.[32] Forms have come together, used in various combinations; synthesized, variant methods appear within a single work. The hybridity openly embraces and invites exploration through intrigue, requiring negotiation, intervention, participation or voyeurism. Viewers experience a co-ordination of text (text being broadly defined so as to include images, sounds, objects), sometimes indeterminate, sometimes non-linear, and often interactive. As a poetic form, language crafted by mind and machine (through code, also a language) predominates, and non-verbal elements also create affect and responses translatable into words. All – or selections from all – the possible elements merge in a shapely way into singularly imparted yet multilayered expression.

This book introduces primary methods; its case studies fall short of representing every method of programming used to make digital poetry via the WWW. Ongoing developments and exploration of previously untapped (cybertextual) technologies abound.

For example, a very recent and seldom practised WWW-oriented technique, known as 'augmented reality' (AR), is poised to make an impact on the genre. In AR, interactive computer-generated projections enhance a user's physical environment. AR in literary practice, in works by Caitlin Fisher and others, merges real and virtual image streams to project poems and tell stories. Fisher's installation *Andromeda*, co-winner of the Vinaròs 4th International Digital Literature Award (2008) and part of 'a larger suite of poems, tabletop theatre, web-viewable and immersive augmented reality fictions', is a journey poem enabled by unique software and a 'custom marker library'.[33] Readers physically explore *Andromeda* – a unique type of pop-up book of poems – while using a videocamera (webcam) connected to a computer running code (ARTag) written by Fisher. Physical movement of the book generates elaborate onscreen text and visuals. Fisher uses AR processes to overlay digital imagery on physical objects (shaped as a book in *Andromeda*). As Borràs observed at the 2009 E-Poetry Festival, Fisher's approach to composition contradicts most examples of digital literature because instead of text and image being transferred from page to screen, they move from screen to page.[34] *Andromeda*, a pioneering work of this type, possesses value for various reasons: for its use of 3D space, for being part book, part installation, for its curiosities, uniqueness of presentation, because of its mix of spoken, animated and written qualities enabled by the specialized (SnapDragonAR) software.[35] Such works entice viewers to become physically involved; their design and content compel us viscerally by altering and by feeding new stimuli into our immediate surroundings. Our literary experience begins to include precisely where we are.

As Christine Wilks reports on the netpoetic site, three augmented reality works were installed at the 2010 Electronic Literature Organization conference: Caitlin and Charles Fisher's *Requiem*, which superimposes images and sounds on a card held in front of the screen by viewers; Alexander Mouton and Christian Faur's *Ethereal Landscapes*, which surrounds viewers with an interactive artist's book, generative audio and video; and Amaranth Borsuk and Brad Bouse's *Between Page and Screen*, described by the authors as a 'digital pop-up book'.[36] In addition to its recent arrival in the field, the difficulty of producing AR limits its practice and may curtail its development. The 'mechanical blood-sweat-+-tears AI scripting ...' writes Mez – who incorporates AR into her current project _feralC_[37] – is 'composed of: 1. the construction of functional AR 'voices' (as in workable entity multilogues/actual avatar scripts); 2. ensuring AI/toon/'chARacter' cohesion (assisted by robust/consistent UX); 3. a gruntwork copywriting 'architecture' (construction/maintenance).'[38] As with any work engaging with advanced computer operations, sheer complexity could prevent AR's wider cultivation as a literary possibility.

Another matter to consider, beyond how authors selectively stage progressive artworks, involves bearing in mind the resources from which they draw content, and their reasons for doing so. As detailed below, only six of the case studies (less than one third) consist of materials entirely crafted by the author. Digital poetry always exploits elements of mathematics, computer science and art, but many other conceptual approaches are applicable. Combining files and presenting them via computer screens multiplies possibilities for poetry and, considering the sum or sums of the artistic equation used to distinguish meaning, requires work from the author and the viewer. One liberating and informative poetic concept with respect to building a context for understanding digital poetry emerged in South America over a century ago.

Pero Afonso de Sardinha arrived on the shores of Brazil from Portugal in the mid-sixteenth century to be Bishop of Bahia. Natives in the Aimorés tribe (pagans) ritualistically ate him. This historical event – a spontaneous response to colonial oppression – has been a source of identification for Brazilian artists and used as a foundation for cultivation of heterogeneous expressive forms. Application of this transgressive context has expanded, and has significance and application in today's media environment. Anthropophagy (or cannibalism), the name assigned to this unusual and iconoclastic creative philosophy, was initially announced in Oswald de Andrade's 'Anthropophagy Manifesto' (1928), which proclaims: 'I am only interested in that which is not my own.'[39] External texts and idioms become grist for the anthropophagist's mill, a trait reflected in de Andrade's short poems 'National Library', partially composed of juxtaposed document titles, e.g. 'Brazilian Code of Civil Law/How to Win the Lottery/Public Speaking for Everyone/The Pole in Flames',[40] and

'Advertisement', which adopts the language of advertising copy, e.g. 'All women – deal with Mr. Fagundes/sole distributor/in the United States of Brazil'.[41] In another historical example of anthropophagy in poetry, Raul Bopp's *Cobra Norato*, the brutal hierarchy of the elements in a rain forest establishes a serial poem involving continuous encounters between the elements (e.g. a snake, trees, a river and birds). Bopp favours process rather than destination and engages, emulates and reprocesses natural, conversational sounds, stitching the language of the creatures of the forest into the poem, e.g. '*Tiúg ... Tiúg Tiúg .../Twi. Twi-twi*'.[42] More importantly, Bopp borrows the story of *Cobra Norato* from native mythology, and re-inscribes it in 'very colloquial and popular language'; anthropophagy radiates in its '*ethos* and thinking structures'.[43]

Anthropophagy motivates numerous poets and artists.[44] Connecting anthropophagy and digital poetry promises significant artistic opportunities for a genre that by its nature synthesizes fragments. Critics often discuss the relationship between concrete poetry and digital poetry without considering the cultural imperatives that influenced concrete poetry, including anthropophagy.[45] Augusto de Campos explains in a 2005 interview,

> Oswald [de Andrade] made a distinction between anthropophagy and pure cannibalism – by hunger or by greed – from ritual anthropophagy. Ritual anthropophagy is a branch of anthropophagy in which the cannibal eats his enemy not for greed or for anger but to inherit the qualities of his enemy. The metaphorical, and also in certain aspects philosophical, idea of cultural anthropophagy Oswald promoted was the idea of cannibalizing the high culture from Europe, with the results that one could acquire, or could have from this devourment, and could then construct something really new out of this development.[46]

Transformative expression appropriates given data then warps or reconfigures it to new ends. Such a method certainly corresponds, or perhaps responds, to Dadaist techniques of appropriation, and also relates to the type of cannibalism seen in examples of digital poetry. In anthropophagic texts, authors engage with multiple languages or idioms, devour other texts and icons, and freely remix discrepant methods and philosophical approaches. The cannibalization of texts discovers and re-discovers meaning, establishing alternative perspectives on cultural or personal subjects taken up by authors in the process of textual composition, re-composition and composting. Through anthropophagy, artists freely reshape external influences. This open acknowledgment of plurality makes the concept forever relevant, as an active principle that creates 'difference'.

Another important aspect to de Andrade's conception of anthropophagy imagines changing taboo into totem. De Campos, who began presenting poems on computers in the early 1980s, always rejected the idea that technology is forbidden, and his early engagement with computers was contrary to typical avant-garde methods at the time. As Perloff observes in the early 1990s, 'the most common response to what has been called the digital revolution has been simple rejection';[47] she explains that the consensus amongst most poets was that 'technology … remains, quite simply, the enemy, the locus of commodification and reification against which a 'genuine' poetic discourse must react.'[48] By now, a more realistic perspective sees computer hardware, software and networks as forces capable of presenting vibrant poetic works. Poets, who mainly (historically) considered technology as an enemy, now commit to it as an expressive tool, shifting its identity as something taboo to something totemic.

Digital poetry mints a literary concept by absorbing forms of expression that benefit from production native to digital technology, but manifest in contrast the WWW's primary concerns. Digital poems have inherited the qualities of computer media: poets coura-geously embrace formidable machines, built for the progression of science and business, and these explorations have been fruitful. Assimilation of texts and language unrelated to computer opera-tions endows digital poetry with autonomy. The anthropophagy of early computer poems generated by algorithmic equation reify Modernism's inscription of tentative, nonlinear arrangements of text, sometimes employing use of randomized elements, as in Dada. Instead of computing equations and processing data, we entrust the computer and WWW with creative responsibilities, requiring machines, programs and servers to play a role in the negotiation between author, reader and language. This dynamic reframes what an author is and does, enabling poets to recycle composed texts within new contexts, and to alter the visual materiality of texts in inventive ways.

The anthropophagic aesthetic of mechanically consuming a text to give birth to new text suggests a type of shifting, combined realization. External material is consumed, digested and restated as an altered entity. Historically, this process of absorbing pertinent foreign matter was a technique used to combat and transcend colonialism. Beyond that objective, it instils cultural relevance by promoting the value of diversity and discrepancy in multiple registers. Digital works in the anthropophagic continuum absorb a range of orientations, including analog information, to make intriguing, vibrant expression. Anthropophagic possibility, incorporating a blend of individual expression and structure along with outside elements, contains obvious opportunity for artists, and many poetic interpretations of the anthropophagic analogy (and other artistic engagements with related concepts) have arisen, such as Bill Seaman's notion of 'interauthorship' within his scheme of 'recombinant' poetics, which acknowledges an 'expanded linguistics of interpenetrated fields of meaning'.[49]

Like Concrete poets, digital poets on the WWW approach anthropophagy in distinct and profound ways: (1) through *transcreation*, which processes and redefines the language of 'original' writings; (2) through direct incorporation of external elements, including multiple languages, images and symbols, in the generation of original expression; and (3) in the mechanical presentation of the work, inventing new technological/navigational structures and appropriating coding language. Each of these methods potentially advances a poem into a realm of heterogeneity. While some of these traits are found in print poetry, digital multimedia works supremely support anthropophagic mechanisms which, as Charles Bernstein writes, give us 'a way to deal with that which is external … by eating that which is outside, ingesting it so that it becomes a part of you, it ceases to be external.'[50] An evolving, transitory art, instigated across a century of possibility, emerges with intent, aesthetic polemic and political depth. In a world of just globalization, artists absorb, through consumption, to become another. To transform, one must be transformed. Incorporating anthropophagic conditions into progressive creative schemes is not compulsory. Doing so, however, opens up new promise for the synthesis of discrepant cultures and expressive histories.

In the abstract and in practice, creative cannibalism in digital poetry considers the alphabet's content as well as words and phrases made by letters as potential fuel. Communication of any sort relies on perpetual permutation and assurance of the finite components of language. Recycling or scrambling words can be a critique of the appropriated language, as in (but not only in) Dada. As DJ Spooky (Paul D. Miller) writes, 'The mix breaks free of the old associations. New contexts from the old. The script gets flipped. The languages evolve and learn to speak in new forms, new thoughts.'[51] Ripe with possibilities for reading digital poetic practice, works discussed in the case studies embrace anthropophagy on aesthetic and technological registers.[52] Text-generating computer programs and other calculated methods have processed and automatically permuted databases of words into poems for a half-century, and in recent years the potential content and media-enabling cannibalistic approaches to creativity have expanded wildly with the growth and capabilities of the WWW.

Transcreation, the translation theory that involves imaginatively recreating a concept from one language into another, usefully contributes to discussions regarding contemporary digital poetics. An inventive process of passing between two languages, transcreation occurs within the multimodalities enabled by machine and network processes. As writers transform poetry digitally, they redefine poetics due to media expansion, and further complicate expression and reception by employing bold and creative practices that appropriate materials stemming from sophisticated origins, handled by elaborate machinery.

In his article 'De Campos Thou Art Translated (Knot)', Bernstein explains that in practising transcreation, 'the poet (cannibalistically) creates an original work in his or her right, one no longer beholden to the source.'[53] This characteristic brings about conditions wherein intertextual processes and connections occur in subjective relation to one another. Authors apply multifaceted, shifting filters – sometimes automatically, sometimes interactively – to a work's core, often based on intricate layering of text, dense with meaning, if not abstraction. Thus results of transcreated texts take a further step towards complexity. Creative cannibalism invites difficulty precisely because the input, source materials or foundations upon which authors build works are provocative – their richness challenges and inspires an author to work with them. A simple permutation of language, such as was popular in past iterations of generative forms, was digestible; an author who uses amalgamated, worldly input, including global databases containing both image and language – as in *dbCinema* and *Synonymovie* – does not simplify the process of reading, absorbing and understanding processed text. Transcreation is administered in many works, and its concept perfectly describes procedures found within the types of polyartistry often practised by digital poets.

For Bernstein, transcreation becomes 'a metaphor for refusing dependency'.[54] This perspective permits the artist or writer to be at once dependent on something (an input text, technology or thought) while resisting it as a uniform or static proposition. Transcreators take information (language), and what (as well as how it) is encoded, using them variably, purposefully, as 'a means of appropriating and remaking in one's own right'.[55] Transcreation means adaptation, sometimes involving media. Practised by authors performing trans-formative acts, its effects can be extremely disfiguring; even faint echoes of the input, in some examples, may not emerge.

Artists working in the field at present employ two major identifiable approaches to creative cannibalism. One of these closely relates to historical practices in the field, and the other forges new territory. With regard to text generators, the latest wave of programs has impressively sought to simulate the tones and techniques of specific writers, artists and devices, e.g. Andrews' appropriation of the *Asteroids* motif. Secondly, and more profoundly, digital poets have begun to construct works by using feeds and information directly from the Internet.

Although approaches to the task vary, the foremost cannibalizing characteristic of automatically generated poems and other interactive works is permutation. A finite set of words is used – consumed, processed, regurgitated – again and again. If programming instructions are complex, and the databases employed are capacious, appropriate and effectively applied, the obviousness of such traits can be diminished. Virtual poetry or writing machines can be entirely original, interactive – readers set constraints and parameters, and add vocabulary – and/or seek to simulate a certain style of writing, or the tone of a particular writer. In fact the inclination to incorporate words and verbal intonations used by historically known poets, giving them new context and visibility, is a fascinating attribute and approach to production of the literary text that persists, if not increases, as years pass. In the pre-WWW period, programmers typically rendered a particular poetic form, such as haiku or sonnet, rather than emulate an author. Recent works such as Carpenter's *Erika*, as well as Niss' *The Electronic Muse*[56] and *The Shannonizer*,[57] indicate a desire for the virtual embodiment of a particular author's expressive style.[58]

Output of *Erika* and these other programs formally compares to works made before the WWW launched. The 'voices' of the writers embedded into the program may not always emerge from the noise, but *Erika*'s poetry effectively repurposes the words of writers at its core, Dickinson and Conrad. Other projects may not in the end advance beyond *Erika*, yet pertain to this discussion. Niss' generator uses the styles of contemporary and canonical writers. *The Electronic Muse* creates and accumulates lines written in the manner of Hollander, e.g. 'Since broken and indivisible poverty rages, that works raging possible telephones on dense leaves', Shakespeare, e.g. 'We with speed crept around pale dispriz'd vows', Sexton, e.g. 'If you urinate, then pine-paneled and well-born kisses menstruate', Mullen, e.g. 'Recyclable, we deliver with in the family way lights', Browning, e.g. 'Have vulgarest and apprised word like this natural true ear!', or 'Dick and Jane', e.g. 'Don't play blue ball as good as those green and green dogs now'.[59] While containing fewer complexities than Carpenter's work in terms of variety, users can add vocabulary to the poem as parts of speech, e.g. noun, adjective, adverb, and through the browser edit generated lines. Output featuring the styles of the attributed authors (as above) emerges. *The Shannonizer* is 'a web toy with delusions of literacy' that gives users the opportunity to 'rewrite' input texts.[60] Built on the premises of Claude Shannon's information theory, the program supplies 'editors' named Lewis Carroll, Raymond Chandler, God, Miss Manners, Edgar Allan Poe, Dr Seuss, Hunter S. Thompson and Mark Twain. Here the results of the programmatic processing are not only effective, but humorous. Lewis Carroll's interpretation of the introduction to one of my recent course syllabuses provides a good example:

Digital poetry is a most provoking thing when a most provoking thing when he knows it came whiffling through the past five decades. Poems that grand? Lend her hands. What tremendously easy riddles you ask! Callooh! Don't stand chattering to build an array of flame, and sonically based in his precious nose, because he went galumphing back. Indeed, in their works produced by poets and evaluate virtues of flame, graphical artistry, if you ask![61]

As formulaic as it may be, the new text playfully reflects various aspects of Carroll's style synthetically addressing course materials. Programmers have indicated a desire for the virtual embodiment of known forms, styles, even authors. May we consider a scenario where the dead poets or artists might somehow, at least textually, come back to life via a computer network, bringing collaborative, cyborgian verses to wanting readers? What an odd, but yet not so far-fetched possibility.[62]

On the second register, beginning with the appearance of works such as *Googlism* and *Google Poem Generator*, poets increasingly use Internet feeds to produce immediate works cannibalistically combining intention and randomness. As explained, contemporary mechanisms enable users to create unique poetic texts made from data instantaneously acquired via the web. Practices tangentially related to these methods have been embraced and exaggerated in the digital-analog writings of the Flarf poets; artists working in this writing collective have become well known for their cannibalistic approach to composition.[63] Flarf methods include using Google search queries to fashion poems fit for print, i.e. they do not adorn words with pictures, animations or other media elements.

One can see the history of interactive digital poetry as developing in three distinct phases. The first involved computation of ASCII text, in which graphical elements were occasionally inscribed; the second brought hypermedia works in many synthetic (fused) forms. We are now at the outset of the third stage, in which, in addition to using previous approaches, authors use the network – the WWW itself – to propel composition.

One way or another, many digital poems cannibalize; in fact the genre's first poem, 'Stochastic Text' by Theo Lutz, was infused with words and phrases taken from a novel by Franz Kafka. Anthropophagic tendencies may be tracked throughout digital poetry's existence – a range of historical texts portray the characteristic.[64] The first indications of this trend on the WWW arrived a decade ago, with the appearance of Stefans' *the dreamlife of letters*. Stefans appropriates textual content in *the dreamlife of letters* entirely from a text by Rachel Blau DuPlessis (part of a POETICS listserver roundtable discussion on sexuality), which he then alphabetizes to make a series of 'concrete' poems based on the chance meeting of words.[65] Stefans' work, as a Flash animation, refines the aesthetic presented in earlier works: words twist upwards in spirals, spin like a propeller, stack into grids and rows, bounce, flowing in vertical columns, and blend into one another. As Janez Strehovec observes in 'Text as Loop', Stefans creates 'meaning' in the poem through the 'quick transitions to anti-words, derivative words, and even non-words'.[66] Rearranged words become something else when put into motion, differing from their performance on a page. Visually compelling, rapid presentation of asyntactic fragments, keeps readers attentive to the matter of their interconnectivity.

Of the two dozen case studies presented above, six works can be observed as containing, in their entirety, material completely original to the title: Larsen's *Carving in Possibilities*, Sondheim's 'Tao' and 'Dawn,' Bouchardon's *Loss of Grasp*, Flanagan's *[theHouse]* and Mez's *ID_Xor-cism*. Otherwise authors inscribe cannibalistic sensibilities, in various ways, into the artistic works introduced. aND's remix of Menezes' name and Higgins' poems present mechanical-spiritual artistic tributes. *Stir Fry Texts*' title may have intended to imply something that was eating, or being eaten; no less than two of its sections utilize appropriated quotations. *Arteroids* feeds several borrowed default texts into the output, also allowing viewers to add text. *Wotclock*, as explained, cannibalizes the text of a previous peripatetic time-based project titled *What We Will*. Rosenberg's process of cultivating text from his own notes and poems, summarized above, involves subjecting written passages to various treatments, finally feeding into his *Diagram* poems. Sondheim, in his *Internet Text* and Second Life productions includes regenerated materials, and extensively appropriates code in order to create his avatars and environments in Second Life. A section of Bouchardon's *Touch* is built atop (and thematically relates to) texts composed by

Aristotle. In Ferraiolo's poems, the media elements in *The End of Capitalism* cannibalize film and video archives; *Map of a Future War* not so subtly gathers and contextualizes corporate and scientific iconography. Sérandour, alternatively, exploits Internet protocols and common tasks to create his poem. Jason Nelson's online poetry game *I made this. you play this. we are enemies* features 'levels designed/destroyed around' the interfaces of numerous well-known websites, and is possibly made with code appropriated from another source.[67] Lehto's *Google Poem Generator*, Tisselli's work, as well as Andrews' *dbCinema* and Zellen's *Without a Trace*, overtly cannibalize the network, feeding off WWW-based resources and information in order to construct digital poems. Poets collaborating with the WWW can refine the raw data presented by its mechanisms. Information's utility shifts from its initial intent and its cannibals exhibit how acquired content can be purposeful beyond original functions it may have held.

Poets, and digital poets, use language as a conduit – a structure capable of containing multiple registers. In these examples we see intake and transformation of information, recycling of text by programmatic configuration into variant formations. In comparison to historical predecessors, advancements abound. Andrews' *dbCinema*, for example, enables much spontaneous intervention by the user, making the output more customizable. Generators, too, have become more comprehensive and continue to expand the particulars of human input and control. Do authors and texts achieve anthropophagy in these works, a synthesis of discrepant cultures and expressive histories? Perhaps not exactly; in these examples we see – especially in the expanded, network-based works – great variation as a result of harnessing to the WWW and instantiation of pre-existing works. Andrews, Lehto, Tisselli and Zellen, ahead of the curve, lay the foundation for future refined works that automatically draw contents from the WWW. Works like Carpenter's show that a programmer can also effectively control and give voice in a versatile way to a constrained textual database. Manipulating larger archives poses more difficulty, and mastering the processes of crafting will develop over time. Since text-based poetry generators took nearly fifty years to reach the successes of *Erika*, we can expect the network synthesizers to take years to advance. Methods of digesting and reordering the quantity of information available will be cultivated, strengthening the powers of author, user and potential content.

Poetry – whether in digital or printed forms – is a transformative act. Through language, now combined with media and sometimes driven by interactivity, an author's responsibility involves setting up structures that manage to 'transport' the reader/viewer to a different place and mindset, or to help them see something in the world from a new perspective. While I would not dedicate or limit artistic methodology to a single compositional approach, processes of cyborgian, networked authoring, using chance and intention, while significantly working within the structure of a computer program, presents interesting challenges and possibilities to poets open to collaborating with a machine, the results of which can be endlessly condensed or expanded though editorial procedure or other forms of 'mash-up'. Cannibalistic artworks seek and gather related information, in order to convey them – through automated and mental activity – into unexpected verbal-visual-sonic interactive pieces. Feeding back (or forward) often drives the text, as creative cannibalism engendered by authors forges content.

These attributes force us to explore and establish new modes of reading and creating contexts for works. At E-Poetry 2009, several scholarly papers referred to examples of cannibalistic digital art and poetry with some new takes on the subject, giving credence to the concept as a type of historical awareness and strategy within the genre. Roberto Simanowski likened the idea of cannibalism to remediation and considered its affinities with postcolonial studies (or anti-colonialist strategies), exploring the concept as a reaction to 'xenophobic movements'.[68] Simanowski addressed 'the other' in digital media and how it is devoured, focusing on how text can be regurgitated as a visual object, as sound, and performance, sometimes stripped of original linguistic content.[69] Bootz made the point that notions of cannibalism are not limited to text but apply to technology itself – our machines, man-made forces of computing and digital media, devour us, and the 'real' cannibalism involves this consumptive relationship.[70] Markku Eskelinen also expressed interest in the anthropophagy between text and user, imagining a scenario when the text takes actual physical information from the user, i.e. measures bodily states and does something with them. Amy Sara Carroll, in her comments on cannibalism, stated that the question for her (along with her collaborator, Ricardo Dominguez) resists addressing co-optation, but rather focuses on the relationship between the word 'cannibal' and Caliban, thereby returning the concept to Postcolonial theory and its relation to Latin America. Carroll read a mashup of texts she had written previously, combining them with an animation comprised of a pastiche of quotes and comments (including Concrete poems by de Campos), while Dominguez projected some of these texts, animated, onto the audience, floor and furniture in the room. Their presentation was a hybrid performance – a combination of imagery, of horizontal (prose) cultural commentary, and critique with (vertical) lines of poetry and poetic references.[71]

Across the gamut of network-based productions presented at E-Poetry 2009, disparate cannibalistic traits could be identified. Digital poets use plentiful methods to reclaim and reuse online voices, words, and images of other artists to creative ends. Christine Wilks' presentation of the website *Remixworx* divulged a particularly impressive display of cannibalism-by-design. Wilks showed a series of animations posted on this collaborative blog, where members of the group have produced hundreds of multimedia remixes since 2006. A member of the group presents an artistic work, which other members of the collective re-make (respond, repurpose) then post in subsequent comment fields, so as to reflect the evolution of the original. Beyond the high quality of the artworks, the collaborative axis of *Remixworx* commands respect, and the sheer variety of types of works (stylistically/aesthetically) embraced by the collective – usually involving kinetic visual poems in combination with graphical animation and sound – is remarkable.

Patricia Tomaszek presented an interactive WWW-based audio piece, a performance tool called 'about nothing, places, memories and thoughts'. In Tomaszek's work, users combine audio samples of lines of poetry by Tomaszek and Robert Creeley, chosen by Tomaszek, making new (mashup) poems by fusing their words. Rettberg's long poem *Frequency* associates each line of his composition with an automatically downloaded (Creative Commons) image from the Flickr photo sharing website that embeds one of the site's 100 most popular tags. Jörg Piringer used Twitter feeds with the word 'Haydn' in them as the basis for the visible verbal element in one of his pieces: letters that appeared were then 'sung' by cut-up pieces of vocal samples of Haydn's compositions. In *Scriptor*, Stefans draws text from the *New York Times*, which his program converts into vibrating, segmented letters that form and reform, (sometimes into words). That Stefans chooses to animate dynamically every point and line of a letter from a text about Barack Obama and Afganistan was a particularly significant example of anthropophagic enterprise in which the voice of the dominant culture undergoes creative transformation. We begin to see language, words – at the levels of letter, line, syllable and sound byte – and images used in mass media perform in a way that ushers in a new awareness of the conditions of speech.

Creative cannibalism is not only pursued, but is pervasive.[72] Computer networks, software and programming engineer transformative experience using collected materials, and authorial approaches expand and diverge from the initial inclination to recycle materials contained in a stand-alone database. Digital poems reflect and rely in various ways on collaborative use of network technology. Are these digital literary experiments examples of anthropophagic texts? Yes: in all instances, readers confront surprising, technologically driven recombination drawn from exterior sources. Some of the remixes teach us about the world we live in, others about how the languages we use to communicate function. Contemplative or expressive states that manipulate or reorganize already unforeseen circumstances become part of a viewer's receptive experience. A direct message might remain elusive but the representation of information is purposeful: multiple forces implement an aggregate entity made by fragments, a concept predominantly forsaken in the hegemonic world.

Leaving off here, intimating that textual and graphical cannibalism signifies digital poetry's strongest trait, not indicating it as simply one possibility among many within digital poetics, would be unfair to countless bona fide lenses applicable to even these very same works. What form of progress can be represented by something defined homogeneously in the twenty-first century? Critics of digital poetry busily labour to define the genre from varied perspectives, teaching the world how works of electronic writers today forecast new literary forms, conceptions of literacy and literary modalities.

Many reasonable assessments, coming from many different directions, explain how these works represent contemporary capabilities and trends in expressive utterance, which often elude wholeness and a sense of completion because those finalities are impossible to achieve. Works resisting finality cannot be easily catalogued – such incredibly varied and versatile forms of expression demand consideration from a multitude of perspectives.

Without question a place for experiments and experiences that look to the arts as a synthesis of other arts, including poetry, has developed. As attempts to define (and confine) electronic/digital poetry continue, creative versatility grows alongside technological advancements in communication. In fact, because the WWW makes it so possible, I would like to see more cultivation of group efforts. As divulged in *Prehistoric Digital Poetry*, informal group efforts by artists working together with software (TRAVESTY, Diastext) and hypermedia (*Alire*) unquestionably accelerated progress in digital poetry. Using communal space with purpose – and collaboration in general – would still appear to be an appealing direction for the field, as evidenced in the formation of works in Second Life and by efforts by participants in *Remixworx* (above); at least two group blogs (Netpoetic,[73] Netartery[74]) now devote bandwidth to discussing developments, stimulating progress and connecting working artists and researchers in the field.

Artists introduced in this book use computer science and technology to explore or expand literary reality, and make poignant pronouncements regarding our tormented, tenterhook world. Amid these staged productions, important applications of ideas aim to stimulate audiences. With heightened standards and capabilities, making art with the aid of a computer is no longer and not in itself reason for celebration. Factors contributing to the success of works include effective use of time: striking a balance between making viewers familiar – yet not necessarily comfortable – with the process while leading them along in a manner that allows for tension and contemplation. Infusing projects with familiar unfamiliarity and discrepancies – as in cannibalizing works – offers the benefits of surprise, impart a lack of predictability, and perpetually perform for the audience.

Digital poetry on the WWW grows and expands, not in a unified direction but pluralistically. Works presenting difficulty to viewers risk marginalization; authors who attend to such practices rely on a presumption that readers or viewers will spend time with sophisticated expression, even if that flies in the face of the WWW's overall identity of providing information quickly and selling products. Elsewhere across its wide spectrum, digital poetry features increased cinematic properties, an unsurprising progression for a field existing alongside cultures becoming increasingly mediated (televisual) in recent decades, even if non-interactive work, as Memmott writes, 'undermines the expectation of hypermediated poetry' and does not take full advantage of the special properties of the WWW.[75] At the end of *Prehistoric Digital Poetry* I observe that given evidence presented further attention to interactivity 'will elevate the genre's profile'.[76] Many of the artists examined, e.g. Andrews, Larsen, Bouchardon, Carpenter, Ferraiolo, Flanagan, Nelson and Tisselli, demonstrate interactivity as a dominant feature in digital poems since the dawning of the WWW, but whether this raises the genre's visibility beyond its particular subculture is unknown. Interactivity requires finite reader involvement with texts, provides multiple sets of possibility and diversity within a given set of texts, and enables exhibition of assorted layers of multimedia. Does it make understanding meaning more difficult? Or are these effects used as bait to pull the reader into something they would not normally find interesting? Getting used to reading complex arrangements of digital literature requires time and effort; less challenging works, as a result of changes in mass literacy due to television, WWW, SMS etc., do provide a way to make literary artworks accessible and more interesting for a public uninterested in reading demanding literature. Someone might also 'find' 'poetry' because of its availability in contemporary digital formats and forms. Developing participatory, interactive mechanisms that surrender control of the text to the reader/viewer would seem to enliven and personalize the experience of the digital poem, and such works have become easier to make and have appeared on the WWW during the past decade. Whether or not audiences locate and enjoy these artworks remains uncertain, although since the field continues to grow and expand, presumably interest, if not demand, exists on some scale.

In electronic, digital, network space, it should be presumed that art will be different. At its most effective it may have the aura of, and may even look like, haiku – a very swift presentation containing very diverse resources. Since we live in an era when trailers for films may be better than the film themselves, literary-minded artists have fantastic opportunity to register in an audience's consciousness because when the Internet functions properly everything appears swiftly, and a potentially compressed art could be preferable. Yet it would be unfair conjecture to posit subtle works of greater duration as inherently less successful than those rapidly radiating a message. Within the vast range of expressive human communication, attention-grabbing texts may garner more attention, although more demanding works could, as a result of the sophisticated development of contents, be critically considered stronger efforts.

Contemporary works feature unstable and precarious compositional designs, in which relations between textual components have evolved. Strehovec's essay 'The Poetics of Elevator Pitch' addresses electronic poetry in the age of short attention spans. Digital poetry, as a textual, meta-textual, linguistic and sometimes non-linguistic practice, requires extended forms of perception. But because human observational skills have changed, Strehovec proclaims the importance of first impressions, getting viewers excited and immediately involved with language. He promotes the notion of an 'elevator pitch' as a temporal ideal for digital poetry – the idea that the poem, 'can be delivered in the time of an elevator ride, e.g. thirty seconds or 100–150 words)', which 'hooks the reader/user within a very short temporal unit.'[77] I take his position seriously because digital writing always appears as a series of shifting mosaic passages, both pointing to potential for instability at any moment and acknowledging unexpected possibilities that happen over time. An everyday vehicle can bring someone to unfamiliar places, and sometimes we are forced to encounter problems caused by something beyond our control. Actions or seductions, in physical or virtual space, immediately and over time, alter social and communicative interaction; the unknown and unexpected are valued in digital writing.

Strehovec's position relates to concerns of literary manufacture because something presented quickly can connect with ideas of permanent and universal interest, can be worthy of being remembered, can contain beauty in expression and form, have intellectual and emotional appeal, and very much represents this particular period and culture. The power of surprise, spectacle, seduction and irregularity provide a solid basis for discussion on digital poetry and poetics. Digital poets encounter simultaneously two major obstacles: they are challenged to satisfy literary/poetic expectations and at the same time confront the manner in which people ordinarily use media, which imposes hindrance on those seeking depth.

Strehovec recommends striving 'for the novel and sophisticated procedures'.[78] Who would argue with that as the momentum of both Modern and Postmodern poetics and artifice extends through electronic literature? Broadly speaking, many authors understand the value of having strong if not sensational graphical appearance at the outset, and, logically, present something surprising or stimulating (pleasant or other) in a work's initial phases. Immediately, and over time, direct messages may remain elusive as mediated expression transmits effects communicating in ways that instil different types of responses, often based on its audience's ability, and openness, to receive the work. Collecting fragments to make wholes requires effort from any reader; those unprepared to do so may find the experience fruitless. And since we do not step into the same river every time we encounter digital poetry, a step-by-step set of instruction resists practical implementation. Enormous varieties in method and form exist. Reflect, for a moment, on the extreme differences between works discussed in the case studies: few of the works directly compare. This fact complicates the ability to provide tactical – compositional and receptive – generalities establishing which approaches work best. Dan Waber's series of brief, fluid animations in 'Strings' provide practically instant gratification, illustration and point for a skimming reader; Cayley's *wotclock* perplexes in its elongated, unconventional delivery. Both indicate legitimate routes for digital literature.

Proponents of Strehovec's line of thinking would quickly argue continuous attention span, or the amount of time a human can focus on an object without any lapse at all, is very brief – some research indicates 'focused attention' may be as short as 8 seconds.[79] After this amount of time, our eyes tend to shift focus, or stray thoughts briefly enter consciousness – though these minimally distracting, momentary lapses do not tend to interfere with task performance, and humans are said to have sustained attention spans of twenty minutes. This does not mean our literature has to be relegated to such brevity, although it certainly explains why services such as Twitter, not to mention sound bytes, garner popularity and digital poetry does not. Audiences having to register and remember, to build narrative or a concept, a barrage of communication within digital poetry, which often exceeds the absorptive capacity of even the most attentive viewers, is sometimes precisely the point. Literature in digital formats continues to play a role in expanding cultural awareness by presenting such challenges.

Hayles identifies two distinct cognitive modes to describe variance in shifting reading practices – if not dimensions of literacy – past to present. In the past, reading in humanities disciplines required 'deep attention … characterized by a willingness to spend long hours with a single artifact'.[80] Today's reading calls for 'hyper attention,' which she characterizes as 'a craving for continuously varying stimuli, a low threshold for boredom, the ability to process multiple information streams simultaneously, and a quick intuitive grasp of algorithmic procedures.'[81] Hayles observes that hypermedia environments create 'hyper attention', a condition enabling viewers to make conceptual connections across a wide variety of domains. 'Students today don't start in deep attention. They start in hyper attention,' David Glenn quotes Hayles as saying in *The Chronicle of Higher Education*, 'and our pedagogical challenge will be to combine hyper attention with deep attention and to cultivate both.'[82] In general, these modes decently categorize historical and contemporary expectations on readers, but they should not be considered as mutually exclusive. Works such as Cayley's *wotclock* (which passes slowly over time), Rosenberg's Diagrams (which also take hours and multiple viewings to absorb), and other contemplative titles (e.g. *Carving in Possibilities*, 'Tao') demand considerable attention from viewers. Evidence presented in the case studies suggests digital poetry usually requires 'deep attention' in addition to 'hyper attention'. Ephemeral works such as *Mesostics for Dick Higgins* and *Arteroids* rapidly pass then disappear, and the process of deeply attending to and brooding over materials may be perverse. Serious viewers, nonetheless, need to find a way to capture and consider the content beyond the moment of its existence, to make the work reverberate while and after it passes.[83] In digital poems, information progressively streaming across screens resists simplification. Poetic pursuits discussed in this book, even those that demand 'hyper attention' (e.g. aND, Andrews), carefully embed detailed information; readers who really want to know and understand complex titles must consider more than what appears on the screen. Attaining deep attention may not always be central to the process of absorption, but certainly may play a role in fully appreciating what transpires in digital poems. Due to this aspect of the work – complex narrative and presentational strategies – digital poetry may always sequester itself on the margins as a result of its virtues. Perhaps only scholars will rigorously attend to works expecting 'hyper' attention of the readers, and doing so may be counterintuitive to the reception of the work. Even so, most works require a combination of 'deep' and 'hyper' attention – attention doubled. Even with deep attention to the surface materials, oversight may occur; without considering other textual variables, such as code and intertextual considerations, realms which establish meaning may be overlooked.

In celebrating the amalgamation of opposing forms of attention, Hayles is spot on, equally condoning Strehovec's inclinations and acknowledging other effective possibilities. Artistically speaking, a more important issue becomes the matter of encountering art that loses our attention. Constantly refreshing attention, relentlessly seducing the reader through variation and textual and aesthetic diversity – operative traits witnessed in works such as Nelson's *I made this. you play this. we are enemies*, Bouchardon's *Touch*, Ferraiolo's *The End of Capitalism* and Andrews's *dbCinema* – have been key elements contributing to the success of these and other works. Despite the possibility that dynamic computer environments may decrease the attention span, and may reduce the ability to focus on a single topic, since new texts and information are easily transferred, digital poets need to be cognizant of computer users who have media literacy but not literary literacy. Given the attention span and sometimes temporal constraints of the average WWW or mobile device user, artists may benefit from making works that do not fluster, and can be read in small chunks. Further, authors may gain advantage by taking into account that much of the potential audience may absorb content on smaller screens, such as mobile phones. Content layered for profundity, depth or viewer transformation on the desktop may not effectively translate to a mobile device. Someone who wants a snack may not appreciate the offering of a four course meal. Given this scenario, the 'elevator pitch' poetics makes much sense. With so little time to deliver perceptive content, the value of immediate insight, imagistic impression (verbal and/or visual) and reflection can not be understated. Grand musings, elegant descriptions, deep development over time and variable metaphorical exposures may, for certain users, be reliquaries from literatures and poetries past.

The authorial ability to shock by media, stun by visual beauty, avoid boredom in and through language cannot be underestimated. What can happen in the first moments of a digital poem? What other techniques support effective staging in subsequent moments and layers of text that declaim sensibility, meaning, direction, purpose? With these questions, we arrive at the literary dynamics of the digital. A test for any digital poem proposed by Memmott, borrowing his line of inquiry from Artaud, is whether a work has 'the ability to *cause thinking*'; the practice reaches its 'greatest potential' in poems accomplishing this feat.[84] Works discussed in the case studies each individually incite arrays of thought, and also do so as a whole despite being formally distinguishable from one another. Bytes, or brief trans-missions, have to be part of the norm, but can be extended by the rigorous development of expression, moving from media spectacle to a mosaic of layers; these layers of discovery further elevate digital expression into something more widely recognized as literature with relevance to the mainstream readership. Even if imparting depth defeats the purpose of the medium, the interfacial goal can aspire to create immediate appeal and seduction of the viewer in conjunction with creating a need or desire to revisit or explore more deeply. While exceptions may exist, in my view the best digital works absorb our attention from the start, and then somehow move us to want to proceed beyond the first impressions.

What happens at the outset of a digital poem may or may not be easily digested or engender intrigue. A balance of immediate and prolonged stimulation may be presented simultaneously. Opposing a straightforward approach, many works unquestionably engage oblique strategies, but as digital poems layer and extend they add poetic depth, and may greatly appeal to someone whose sensibilities have been trained to appreciate discursion. Digital poets can create works using the best attributes of technology and the most effective literary vehicles. Does it matter if these productions embody sophistication and richness using the principles of great literary works? No – because they maintain and deserve our interest for different reasons, particularly inasmuch as they illustrate neoteric expressive (and receptive) potential, representing advanced and innovative ways to write poetry as a form of communication united with graphical media.

We cannot expect digital poems to find a way to enter mainstream consciousness soon, even though many more productions are forthcoming and may be witnessed on a wider scale. If works resist 'elevator pitch' poetics, which might broaden their appeal, so be it. I recall being taught through writing (and through reading), through fiction and poetry, about capacity *and* truncation, and a variable formula mixing the two holds much promise for digital poetry. I can want either, or both. Tactics and techniques of surprise and seduction over the short and long term can purposefully construct and scale to the marvellous. In conventional poetics, this corresponds with Allen Ginsberg's fundamental decree: 'Maximum information, minimum number of syllables.'[85] Digital authors, through skilled and elegant thinking, may issue plenitude with sparse materials, whether readers need a moment or an afternoon to negotiate and comprehend a work's implications. A bigger critical concern is the effect of doing a little with a lot.

Nathan Brown's 'The Function of Digital Poetry at the Present Time' outlines various problems faced by digital poetry: 'How to unleash the latent dynamism of the object or to manifest its brooding opacity. How to punctuate the aleatory play of the event or to zoom in on its glacial unfolding. How to wage the same war ... against the complicity of banality and distraction.'[86] Brown's concerns warrant appreciation and response, particularly since another view might see positive attributes to artistic distraction and being complicit in banality – that these characteristics might, in the short term, have some benefit. Brown's essay is provocative, even if he ends up arguing for coherence or at least clarity. Now, while I do not wish to promote banality, I must also accept the possibility that cleverly tapping into the commonplace can actually generate a greater audience; and I see distraction in an e-text (or by an e-text) as legitimate, because it holds potential to lead viewers towards transformation. Digital poetry (and the poetics implicit within it) is a capacious endeavour, an inviting genre that can accommodate a wide range of approaches and predilections. As in print literature, vibrant and vital works thriving on the fringes will complement canonically accepted works; those works that canonize, and those that cohere and seek clarity, cannot be acknowledged as the only worthy tastemakers.

Desire for spectacle and brevity in this cultural epoch cannot be ignored, yet we can also expect, for better or worse, a resistance to this circumstance. Robert Coover, lecturing on hypertext, once said, 'half of the writing is in the linking.'[87] Let's agree that the 'writer' provides landscape for the reader – the writer who, even if a haiku-ist, must stimulate on the surface, and beneath and beyond it. Thus refreshing language and appearance of a work at a reasonable pace helps to captivate an audience. Metaphors may limit, but can also be expansive, suggestive models for our screens and minds. A digital poem may be micro, or micro and macro, or macro or autonomous, may be surgical, composite and certainly imaginary/illusory. In digital poetry, we can have our speculative and hyper-attentive spectacles, and eat them too. What's 'them'? Our spectacles are the words and their mediated cohorts (images, sounds, links) looking to correspond with past and future senses of writing.

Epilogue

I complete *New Directions in Digital Poetry* having recently participated in the 2011 E-Poetry Festival. As a performer, I assembled a
pan-hemispheric musical ensemble named grope uSurp, comprised
of other multimedia artists attending the festival; Lucio Agra, John
Cayley, Stephen Cope, Andrew Klobucar, Siew-wai Kok and Eugenio
Tisselli joined me onstage. One of our pieces, proposed by Cayley,
cannibalized a Neil Young tune and was particularly apt to the
gathering. In one verse Cayley sings, 'I've been flying/through the
code/Words are starving/to loose their hold.'[1]

This sensibility coincidentally echoed observations presented
in my talk at the event, 'Bearing the Fruits of E-Poetry: A Personal
Decennial View', which describes how my 2001 E-Poetry Festival
band '9 way mind' strove to emphasize language in its performance,
whereas the 2011 band's preparations largely focused on using
MIDI technology[2] to compose images and language interactively.
This contrast of preparation and performance, in which the role of
language *per se* has been reimagined and reconfigured, is on the
face of it indicative of the aesthetic transition the field of digital
poetry and performance as a whole has sustained. To be sure, well
beyond my own ensemble, what is clear is that during the period
between the first and most recent E-Poetry festivals, a telling shift in
the prominence of language occurred.

Authors present language – indeed, each work discussed in
this book features words – but verbal components are today fully
balanceable with mediated components. In programmatic compositions, such as those permitted on the WWW, words hungrily invite,
and may require, extra-textual supplementation to thrive. This is to
say that, in this new poetic paradigm, words do not surrender their
power but instead share it with that of other expressive elements,
and reading now happens on multiple registers.

I had anticipated that my attending E-Poetry might bring apprehension, might usher in a type of crisis as regards my research,

that a new slew of dynamics might be unleashed, deviating beyond contexts I consider for the genre. They did not. Works presented at the Festival featured new applications, tools and skilful, energetic combinations of forms, all of which were very exciting.[3] Yet I left E-Poetry 2011 charged with understanding, with a sense that digital poetry as presented onstage and on the WWW has reached a necessary plateau, a stage replete with exploration and variations of form that test and perpetually re-establish boundaries. Hence I have arrived at a juncture at which the essential points of departure – while variable and interchangeable – are evident and will remain in place as long as the technologies supporting them are upheld. Of course, a new phase will inevitably arrive, and creative impulses identified in this study will fold into new dynamics held by shifts in hardware, software and network technology.

What conditions will incite movement away from this particular epoch? Presumably it will include device-based developments, a trend which may have already begun. As portable devices move away from the PC/laptop model, creative applications developed specifically for global satellite networks, made with tools not congruous with those used for WWW production, could change the face of digital poetry. Proprietary delivery mechanisms (i.e. hardware) already prohibit viewing entire swaths of work germane to digital poetry at present (e.g. Apple refuses to enable use of Flash players on iPhones and iPads). New devices may or may not be inclined to incorporate technologies compatible to those used on today's WWW, and the likelihood of tomorrow's devices rendering today's processes and productions obsolete is near certain.

This book records a specific moment in the genre's continuum, when it has arrived on a global, multimedia computer network for the first time. Its ephemerality – knowing digital poetry as such will not be here forever – encourages me to explore what these works consist of and to provide possible approaches for reading them.

Acknowledgements

New Directions in Digital Poetry's best resource has been the extra-ordinary series of E-Poetry festivals directed by Loss Pequeño Glazier (2001–Present) and co-organized in the past by Sandy Baldwin (2003, 2011), John Cayley (2005), Philippe Bootz (2007) and Laura Borràs (2009). Presenting research and artworks at five E-Poetry Festivals enabled monumental advancement and opportunity for me.

Other events I attended and presented at, instrumental to this book, include: Science, Technology, and the Humanities: A New Synthesis (Stevens Institute of Technology); The Network as a Space and Medium for Collaborative Interdisciplinary Art Practice (U of Bergen); Colóquio Internacional de Ciberliteratura (Pontifícia U Católica de São Paulo); In(ter)ventions – Literary Practice At The Edge: A Gathering (Banff Art Centre); The International Conference & Festival of the Electronic Literature Organization (Brown U); Interrupt: Language-Driven Digital Art (Brown U); Codework: National Science Foundation Workshop (West Virginia U); Reading Digital Literature: American-German Conference (Brown U); MITH Digital Dialogues (U Maryland); Poem/Art: Brazilian Concrete Poetry (Yale U); BIOS: the poetics of life in digital media (West Virginia U); Thailand Media Arts Festival (Chulalongkorn U); and Festival Internacional de Linguagem Electrônica (São Paulo).

Several publications circulated my preliminary essays addressing materials contained in this book, including: *Jacket2*; *CiberLetras: Journal of literary criticism and culture*; *Harriet*; *Reality Sandwich*; *Postmodern Culture*; *Electronic Book Review*; *dichtung-digital*; *Literary Art in Digital Performance: Case Studies in New Media Art and Criticism*; *Hermeneia*; *Proceedings of Science, Technology, and the Humanities: A New Synthesis*; *Proceedings of E-Poetry 2007*. I thank the following organizers and editors (who are sometimes both) for their attention to, and respect for, my work: Sandy Baldwin, Ricardo Barreto, Philippe Bootz, Laura Borràs, John Cayley, Lori Emerson, Ed Foster, Susana Haydu, K. David Jackson, Ken Jordan,

Matthew Kirschenbaum, Ellen McCallum, Paula Perissinotto, Somjai Sae-teaw, Scott Rettberg, Francisco J. Ricardo, Lucia Santaella, Roberto Simanowski, Steven Ross Smith, Michelle Taransky, Steven Taylor and Edwin Torres.

A Spring 2009 sabbatical granted by New Jersey Institute of Technology, where I originally presented many of this book's articulations to students, provided supplemental time for its research and composition, as did a 2005–6 Fulbright Scholar appointment at Multimedia University (Malaysia). Courses I taught at Naropa University (Creative Cannibalism & Prehistoric Digital Poetry, 2007) and University of Pennsylvania (Digital Poetry, 2010) also informed the writing and positioning of New Directions in Digital Poetry. I extend gratitude to supportive colleagues at these institutions, including Norbert Elliot, Fadi Deek, Ahmad Rafi Mohamed Eshaq, Al Filreis, C. W. Khong, Burt Kimmelman, Robert Lynch, Priscilla Nelson and Anne Waldman, as well as their administrative assistants.

Feedback and guidance I received from numerous readers and companions contributed enormously to New Directions in Digital Poetry. Comments by, and conversations with, Francisco J. Ricardo (and other readers at Continuum), Lucio Agra, Jorge Luiz Antonio, Wilton Azevedo, Sandy Baldwin, Charles Bernstein, Friedrich W. Block, Philippe Bootz, Augusto de Campos, J. R. Carpenter, Maria Damon, Lori Emerson, Kevin Fitzgerald, Amy Hufnagel, David Jhave Johnston, Brendan Kiley, Burt Kimmelman, Andrew Klobucar, Talan Memmott, Nick Montfort, Ted Nelson, Marcus Salgado, Roberto Simanowski, Alan Sondheim, Brian Kim Stefans, Rui Torres and André Vallias resonate in this book's chambers. Sophia Sobers helped me prepare the book's diagrams; Davida Scharf provided important technical assistance. To everyone who helped push me forward, not the least of whom are all of the authors who created intriguing works, and engaged in dialogues with me about them, my appreciation!

Finally, and most importantly, personal support I receive from my parents and family, particularly my wife Amy Hufnagel and daughters Constellation and Aleatory, provided the ultimate foundation, impetus and ability for the conduction of my efforts.

Notes

Notes to Introduction

1. Manfredo Tafuri, *Architecture and Utopia: Design and Capitalist Development*, trans. Barbara Luigia La Penta (Cambridge, MA: MIT Press, 1976[1973]), 83.

Notes to Chapter 1

1. J.Y. Douglas, 'Understanding the Act of Reading: The Woe Beginner', *Writing on the Edge* 2, no. 2 (1991).
2. My seminars are taught as online courses, where ongoing dialogues transpire by way of virtual conferencing tools such as Moodle.
3. Jonathan Vo, 'Re: Arteroids', (New Jersey Institute of Technology, 2010).
4. Noelle Screen, 'Re: Arteroids', (New Jersey Institute of Technology, 2010).
5. Noah Wardrip-Fruin, 'Reading Digital Literature: Surface, Data, Interaction, and Expressive Processing', in *A Companion to Digital Literary Studies*, ed. S. Schreibman and R.G. Siemens (Malden, MA: Blackwell, 2007) 163.
6. C. Funkhouser, *Prehistoric Digital Poetry: An Archaeology of Forms, 1959–1995* (University of Alabama Press, 2007). Among the ways the previous book defines digital poetry are as follows: 'Digital poetry is a term that represents a spectrum of computerized literary art that can be appreciated in the context of the poetic tradition', 24; 'The genre combines poetic formations with computer processing or processes', 22; 'Digital poetry is a genre that fuses crafted language with new media technology and techniques enabled by such equipment, and is a reasonable label to use in describing forms of literary work that are presented on screens with the assistance of computers and/ or computer programming', 23.

7.   Physically responsive software fictions that forego historical
         conventions, such as use of a computer mouse, have emerged;
         see, Charlotte Davies' *Osmose*, and *The Breathing Wall* (created
         by Kate Pullinger, Stephen Schemat, and babel), which monitor
         a viewer's breath. Other works reflecting aesthetic and technical
         progression and innovation within the genre include Serge
         Bouchardon's digital poems *Touch* and *Loss of Grasp*, with
         which viewers interact through gestures on network-connected
         webcams (see Ch. 4) and Jörg Piringer's vocal-response
         animations, where texts are propelled by sound and the
         speaker's voice.

8.   Works introduced in the MITH talk included Pedro Barbosa's
         SYNTEXT, Stefans' *the dreamlife of letters*, John Cayley's
         riverIsland, mIEKAL aND's *Seedsigns for Philadelpho*.

9.   'History of the World Wide Web,' en.wikipedia.org/wiki/
         History_of_the_world_wide_web.

10.   HTML (HyperText Markup Language), the 'markup language'
         used to describe the structure of text in WWW pages, conceived
         by Tim Berners-Lee in 1991 and developed by the World Wide
         Web Consortium, was standardized in 1995. Markup Languages
         alter the presentation of the content of a document.

11.   Robert Hobbes' Zakon, 'Hobbes' Internet Timeline', www.zakon.
         org/robert/internet/timeline/.

12.   Ibid. When the WWW was launched, the network was capable of
         transmitting information at 'T3' speed, or 44.736 Mbps (megabits
         per second); today, via Internet2, the bandwidth is an extremely
         rapid 10Gbps; Mbps and Gbps are measures of bandwidth, or
         the total information flow across a telecommunications medium/
         network.

13.   Ibid. Despite the mammoth growth in computing, both off- and
         online, Internet Protocols remained the same from 1979 to 2000,
         when Internet2 was launched.

14.   A decade later, the same technology would be used as the basis
         for the type of streaming video (.mpeg-4 or .mp4) used to propel
         popular video sharing websites such as YouTube.

15.   For example, the release of the Netscape 3.0 browser in 1996
         allowed JavaScript to be included in HTML. JavaScript enables
         the composer to embed extra functions within an HTML context,
         including customized 'pop up' windows, input forms that can
         be validated, and other interactive features (such as mouseover
         effects). Netscape also featured a built-in HTML editor, using
         tools similar to those found in word processing programs,

so anyone who wanted to could create or edit basic WWW pages without knowing HTML. Several HTML editors, such as HotDog (1995) and HoTMetaL (1994), became freely available; others, such as Microsoft FrontPage (1996) and Macromedia Dreamweaver (1997), were commercially released. In 1997, DHTML, a browser feature which essentially combines HTML and JavaScript to enable special effects and animation, became available.

16. See *Geeks Bearing Gifts* for a detailed history of the development of Flash.

17. Theodor H. Nelson, *Geeks Bearing Gifts: How the Computer World Got This Way* (Sausalito, CA: Mindful Press: Distributed by Lulu.com, 2008), 17.

18. For a thorough explanation of the components and programming mechanics contained in Flash, see David Sheperd's essay 'Finding and Evaluating the Code', referenced in Chapter 2.

19. Since the turn of the century, other technological and commercial phenomena – sometimes gathered under the rubric 'Web 2.0' – such as the proliferation of blogs, increased search capabilities, virtual worlds, content syndication (RSS), social networking hubs (Facebook), and streaming media sites have also played a role in the distribution and discussion of poetry and digital poetry, sometimes profoundly so (e.g., *dbqp: visualizing poetics*, *Silliman's blog*, YouTube, *PennSound*).

20. 'Object (Computer Science)', en.wikipedia.org/wiki/Object_(computer_science). Object-oriented programming involves using 'objects' that contain 'data and the instructions that operate on those data'; the object is built through the programming instructions.

21. Another notable example, not further explored in this book, is the cultivation of Perl programming language, a powerful text processor that has been uniquely used to compose visual poems using code that can also be compiled by a computer. See 'Perl', en.wikipedia.org/wiki/Perl. A type of 'code poetry' which also functions within the machine is unique to Perl because of the large number of regular English words it uses. A number of Perl poems are published on the website *PerlMonks* ('Perlmonks', www.perlmonks.org/?node=Perl%20Poetry).

22. Archives housing impressive works such as *Ubuweb* (www.ubu.com/) and *Electronic Poetry Center*, (epc.buffalo.edu/) were established at, or near, the outset of the WWW.

23. URLs for these sites are, respectively: D. R. Slattery, 'Alphaweb',

iat.ubalt.edu/guests/alphaweb/; Christy Sheffield Sanford, 'Flowerfall', www.thing.net/~grist/l&d/bpnichol/tta/ch/coup.html and Mark Amerika, 'Grammatron', www.grammatron.com.

24. In *Geeks Bearing Gifts,* Nelson defines a database as 'any principled arrangement of information that you can look things up in', p. 36. Importantly, he observes, 'In the mainframe era, 1947–87, everything was implicitly a database; only gradually did the database become a separate concept', 37.

25. Geof Huth, 'Dbpq: Visualizing Poetics', dbqp.blogspot.com/.

26. Crag Hill, 'Crg Hill's Poetry Scorecard', scorecard.typepad.com/.

27. mIEKAL aND, 'Spidertangle', www.spidertangle.net/home.html.

28. Talan Memmott, 'Beyond Taxonomy', in *New Media Poetics: Contexts, Technotexts, and Theories*, ed. Adalaide Kirby Morris and Thomas Swiss (Cambridge, MA: The MIT P, 2006), 304.

29. Michael Joyce, *Of Two Minds : Hypertext Pedagogy and Poetics* (Ann Arbor: University of Michigan Press, 1995).

30. Espen Aarseth, *Cybertext: Perspectives on Ergodic Literature* (Baltimore: Johns Hopkins U P, 1997). Adopting Aarseth's general paradigm as a starting point makes sense in terms of declaring the most basic dynamics and behaviour of a work. 'In ergodic literature', he writes, 'nontrivial effort is required to allow the reader to traverse the text.' (1). The corollary to this concept is 'nonergodic literature', where 'effort to traverse the text is trivial', as in turning the page of the book, or clicking through a linear text on a screen. Nonergodic, digital poems are linear (non-hypertext); readers participate by observing and by constructing a subjective meaning for the work; as is customary for any Postmodern artwork, authors of digital poems generally supply a context for the materials, and viewers cognitively assemble mediated fragments to gain a sense of a work's purpose. Ergodic literature is explorative: an author presents interlinked texts (hypertext/media) or a virtual object (as explained in *Prehistoric Digital Poetry*, 165); ergodic poems can also be constructive, offering viewers the opportunity to incorporate their own tangible input (making additions/amendments) to a flexible digital structure designed by an author.

31. Stephanie Strickland, 'Writing the Virtual: Eleven Dimensions of E-Poetry', *Leonardo Electronic Almanac* 14, Nos. 5–6 (2006), www.leoalmanac.org/journal/Vol_14/lea_v14_n05-06/sstrickland.asp.

32. Ibid.

33. Ibid.

34. Ibid.

35. Stephanie Strickland, 'Quantum Poetics: Six Thoughts', in *Media Poetry: An International Anthology*, ed. Eduardo Kac (Chicago & London: Intellect Books, 2007).

36. Several relevant foreign-language titles on the subject have appeared since 2008, including: Jorge Luiz Antonio's *Poesia Eletrônica: negociações com os processos digitais* (2008); Serge Bouchardon's *Littérature numérique: le récit interactif*, Camille Paloque-Bergès' *Poétique des codes sur le réseau informatique* (2009): Dionisio Cañas and Carlos Gonzáles Tardón's *Puede un computador escribir un poema de amor?*, and Roberto Simanowski's *Textmaschinen – Kinetische Poesie – Interaktive Installation* (2010). Other anthologies in English devoted to the subject include Philippe Bootz and Sandy Baldwin's *Regards Croisés: Perspectives on Digital Literature* (2010), Francisco J. Ricardo's *Literary Art in Digital Performance: Case Studies in New Media Art and Criticism* (2009), and Peter Gendolla and Jörgen Schäfer's *The Aesthetics of Net Literature: Writing, Reading and Playing in Programmable Media* (2007).

37. N. Katherine Hayles, *Electronic Literature: New Horizons for the Literary* (Notre Dame, Ind.: University of Notre Dame Ward-Phillips, 2008), 2.

38. The *Electronic Literature Collection* Vol. 1 was released on DVD and the WWW in 2007, and included with Hayles' *Electronic Literature* (2008), 2.

39. Hayles, *Electronic Literature: New Horizons for the Literary*. The *Electronic Literature Collection* Vol.1 was first in an ongoing series of such anthologies produced by the Electronic Literature Organization. Vol. 2 in the series, edited by Laura Borràs Castanyer, Memmott, and Rita Raley was issued in 2011.

40. Leslie Mezei, 'Computer Art...,' *Artist and Computer* (1976), www.atariarchives.org/artist/sec7.php.

41. Joyce, *Of Two Minds: Hypertext Pedagogy and Poetics*, 236.

Notes to Chapter 2

1. John Lavagnino, 'Digital and Analog Texts', in *A Companion to Digital Literary Studies*, ed. S. Schreibman and R. G. Siemens (Malden, MA: Blackwell, 2007), 413.

2. Carlos Drummond de Andrade, Rafael Alberti and Mark Strand, *Looking for Poetry: Poems by Carlos Drummond De Andrade and*

Rafael Alberti and Songs from the Quechua (New York: Alfred A. Knopf, 2002), 146.

3. Cited in N. Katherine Hayles, *Electronic Literature: New Horizons for the Literary* (Notre Dame, Ind.: University of Notre Dame Ward-Phillips, 2008), 160.

4. Bruce Andrews, 'Electronic Poetics', in *Cybertext Yearbook 2003*, ed. Markku Eskelinen, et al. (University of Jyväskylä, 2003), 34.

5. Rasula, J. 'From Corset to Podcast: The Question of Poetry Now'. *American Literary History* 21, no. 3 (2009), 660.

6. Cited in Bertrand Gervais, 'Is There a Text on This Screen? Reading in an Era of Hypertextuality', in *A Companion to Digital Literary Studies*, ed. Susan Schreibman and Ray Siemens (Malden, MA: Blackwell, 2007), 185.

7. J. Drucker, 'The Virtual Codex from Page Space to E-Space', in *A Companion to Digital Literary Studies*, ed. Susan Schreibman and Ray Siemens (Malden, MA: Blackwell, 2007), 217.

8. Ibid., 226.

9. Ibid., 229.

10. Ibid., 230.

11. Jim Andrews, 'Stir Frys and Cut Ups', collection.eliterature.org/1/works/andrews__stir_fry_texts/text.html.

12. David Shepard, 'Finding and Evaluating the Code,' newhorizons.eliterature.org/essay.php@id=12.html.

13. Ibid. Declarations of the importance of studying code and algorithms have been asserted by several scholars. For example, in *Digital Poetics* Glazier writes, 'Is such attention to detail necessary? Will one's code always be 'read'? The answer ... is that ... it will not always be read, just as poems may not always be 'read.' Nevertheless, it is not about whether a reader will fathom the intricacies of your code. Rather, the issue here is that HTML coding has material dimension and that writing, as an act, means working within such dimensions', 110. Similar concerns have been addressed by Hayles, who writes in *Electronic Literature*, 'Critics and scholars of digital art and literature should ...properly consider the source code to be part of the work', 35. In the essay 'Reading Digital Literature: Surface, Data, Interaction, and Expressive Processing,' Wardrip-Fruin posits three frameworks for reading digital literature. Most critical work (which would include my own), according to Wardrip-Fruin, 'proceeds from an implicit model that takes audience experience to be primary. The main components of the model are

the surface of the work…and the space of possible interactions with the work', p. 164. The second framework is Espen Aarseth's explicit model, which includes rigorous consideration the work's internal mechanisms. Because 'the implicit audience model provides only a partial view of digital literature', Wardrip-Fruin develops a third model, 'expressive processing', which, after identifying 'the interplay between the work's *data*, *process*, *surface* and possibilities for *interaction*' proposes that 'groupings of theses may be considered as *operational logics* and explored as 'authorial expressions' and as 'expressing otherwise-hidden relationships with our larger society', 165, emphasis Wardrip-Fruin). This differs from the explicit model by not focusing solely on how the algorithm works. In Wardrip-Fruin's scheme, such meta-views are needed because 'the surface may not reveal the aspects of a work that will be most telling for analysis (a case in which scholars may miss what a work's processes express)', 182. Generally speaking, the user of any WWW-based work can use the browser's View menu to glance at the 'Source' code used to create the page. However, while certain aspects of the code (e.g. HTML, JavaScript) are usually readable, the code accompanying digital poems often refers to files (such as Flash, .swf files) whose source cannot be read. Studying source code, as Shepard writes, 'can provide a clearer understanding of the methods behind the program', but this is a task that remains largely outside the purview of this book.

14. Ibid.

15. Hayles, *Electronic Literature: New Horizons for the Literary*, 154.

16. Eventually, as someone begins to appreciate digital poetry, s/he may be inclined to know how code works – perhaps in the way someone who visits and enjoys a country foreign to them may desire or seek to learn its language.

17. While almost anyone can be trained in using a pre-packaged software program, artists and educators who take the initiative to learn a programming language hold distinct advantages over those who do not. The basis of any digital art or literature project is computer code. Developing the skills to manipulate and compose using code enhances both presentational and archival possibilities, encouraging the use of expansive databases that can now be populated with multimedia elements. A designer or producer who interacts with code is more likely to produce, and maintain control over, an original document. Unfortunately, few professors working in English and Humanities departments have taken the time to study and practise working with code. Thus, in

my view, the importance for teachers of digital literature to be practising artists, and for practising digital writers to have some ability to work with code, cannot be understated.

18. Hayles, *Electronic Literature: New Horizons for the Literary*, 152.

19. Ibid., 155.

20. Ibid., x.

21. Coincidentally, Intermedia was also the name given to a significant 1985 hypertext project at Brown University (see 'Intermedia', en.wikipedia.org/wiki/Intermedia_%28hypertext%29).

22. Hayles theorizes technology and the body essentially because, she writes, 'body and machine interact in fluid and dynamic ways that are codetermining.' (128).

23. Hayles, *Electronic Literature: New Horizons for the Literary*, 57.

24. A secondary (dis)connection can be made between Hayles and Higgins in this regard. The subtitle of Hayles' book, 'New Horizons for the Literary', coincidentally shares a common orientation with the title of Higgins' *Horizons: the Poetics and Theory of the Intermedia* (Carbondale, IL: Southern Illinois Univ. Press, 1984).

25. Hayles, *Electronic Literature: New Horizons for the Literary*, 84.

26. Richard Kostelanetz and H. R. Brittain, *A Dictionary of the Avant-Gardes* (New York: Schirmer Books, 2000), 305. Kostelanetz also notes 'Samuel Taylor Coleridge's earlier uses of the epithet.' (305). What Kostelanetz refers to here is Coleridge's coining of the term *intermedium* in a lecture he wrote on Edmund Spenser. In his essay 'FLUXLIST and SILENCE Celebrate Dick Higgins', Ken Friedman writes, 'Coleridge referred to a specific point lodged between two kinds of meaning in the use of an art medium. Coleridge's word "intermedium" was a singular term, used almost as an adjectival noun. In contrast, Higgins' "intermedia" refers to a tendency in the arts that became both a range of art forms and a way of approaching the arts.' (see Ken Friedman, 'Fluxlist and Silence Celebrate Dick Higgins', www.fluxus.org/higgins/ken.htm).

27. E-POETRY: An International Festival of Digital Poetry is a biennial event first organized by Loss Pequeño Glazier at the University of Buffalo-SUNY in 2001. Subsequent events in the series have been held at West Virginia University (2003), in London (2005), Université de Paris 8 (2007), Barcelona (2009) and Buffalo (2011). Primary organizers of the E-POETRY events include Glazier, Baldwin, Bootz and Borràs.

28. Poesia Visiva is a term used to designate a grouping of Italian artists who were involved with the Fluxus movement. My report on E-Poetry 2003, initially published by the Hermeneia research group in 2004 (but presently absent from their website), is available via www.wepress.org/E_Poetry_2003.pdf.

29. Christopher Keep, Tim McLaughlin, and Robin Parmar, 'The Electronic Labyrinth', www2.iath.virginia.edu/elab/.

30. Jessica Pressman, 'Navigating Electronic Literature', newhorizons. eliterature.org/essay.php@id=14.html.

31. Hayles, *Electronic Literature: New Horizons for the Literary*, 10.

32. Marina Corrêa, 'Concrete Poetry as an International Movement Viewed by Augusto De Campos: An Interview', www.scribd.com/ doc/36967268/Concrete-Poetry-as-an-International-Movement-Viewed-by-Augusto-de-Campos.

33. Ibid.

34. Lori Emerson, 'It's Not That, It's Not That, It's Not That: Reading Digital Poetry', *E-Poetry 2003: An International Digital Poetry Festival* (West Virginia U 2003).

35. Hayles, *Electronic Literature: New Horizons for the Literary*, 156.

36. Gervais, 'Is There a Text on This Screen? Reading in an Era of Hypertextuality,' 183–4.

37. John Keats, 'Letter to George and Tom Keats 27/12/1817', in *Selected Letters of John Keats*, ed. Grant F. Scott (Cambridge: Harvard U P, 2002), 193.

38. Brian Kim Stefans, Conversation with the author, May 2009.

39. 'Meme montage' is a term first introduced by Francisco J. Ricardo, in the 'Post Chapter Dialogue' between him and me in *Literary Art in Digital Performance: Case Studies in New Media Art and Criticism*, 80.

Notes to Chapter 3

1. english.umn.edu/joglars/SEEDSIGN/index.html.

2. english.umn.edu/joglars/mesostics/index.html.

3. He is also a farmer who has organized an 'intentional community' in Wisconsin called Dreamtime Village; agriculture has been an important aspect of his life since the 1990s. His *Internalational Dictionary of Neologisms*, which began as a 'mail art' project in the 1980s, was profiled in *Prehistoric Digital Poetry*. In 1995 the *Dictionary* contained 425 entries; aND has transformed this

title into a WWW site to which users can contribute, and as of January 2010 it has 2,782 entries (www.neologisms.us/). aND has created a range of unusual and refined multimedia animations, digital poems and videos, and continues to produce elaborate analog works that benefit from digital methodology. Of particular relevance to this point is aND's collaboration with Maria Damon, *eros/ion*, which started as an online project (2001) and became a hypertextual book containing supplementary commentary. I do not discuss it here because of its largely prosaic appearance, but wish to acknowledge it as a noteworthy literary endeavour. Another important – but no longer accessible – project instigated by aND was 'Writing Debuffets Titles' (2003). 'Writing Debuffets Titles' was a wiki-based collaborative hypertext poem initiated by aND enabling group authoring and editing, which was compromised (and ultimately dismantled because of) uncontrollable spam.

4. english.umn.edu/joglars/multidex.php.

5. Intersign poetry, which also came to be known as *interpoesia* (interpoetry) when computers emerged as a creative force, involves interactive principles, although in addressing the infrastructure in *Poetics and Visuality* Menezes proposes a distinction between 'the standard notion of poetry as "a coded articulation" in which there is "no poetry outside the verbal code", and a related idea that amplifies the first to an "articulation of language"' (np). Being able to understand poetry on both these registers permits 'the creation of poems also from non-verbal signs', a concept not unfamiliar to experimental poetry in Brazil (ibid.). Menezes rejected the concept of limiting the properties of poetic articulations to the verbal realm. Imagery, for Menezes, 'disarticulates the divisions among languages, sectioned up and individualized by the ideological system to reflect the conception of the world' (ibid.), contributing, he writes in a 'Guide for Reading Intersign Poems', 'meaning and comprehension'. (40). *Poetics and Visuality* demands 'a revision in the use of technology', favouring multimedia because its multiple simultaneous channels represent 'a reinvigoration of the parts, disguising them in an experimental mess where each sign is kept within the limits of its semiotic nature; they share the same space but do not blend or fuse' (np). Rather than synthesize disparate elements into a single determination, compounded elements construct various possibilities.

6. mIEKAL aND, ' Seedsigns for Philadelpho', english.umn.edu/ joglars/SEEDSIGN/index.html.

7. Using a scanner, aND arranged a collection of False Blue Indigo

seeds into the letters, which he then arranged into words that comprise the seedsigns.

8. Anagram poems also emerged during the Southern Dynasties in China 1500 years ago.

9. The changes occur (approximately) every half-second in each line, and are staggered so that changes do not happen uniformly in the lines.

10. Words appearing on the second line begin with *d* (deep, dial, deal), the third *l* (loop, load, lead, leap, laid), and the fourth *p* (pool, pile, peel, pale).

11. See Philadelpho Menezes, 'Interactive Poems: Intersign Perspective for Experimental Poetry', www.cyberpoem.com/text/menezes_en.html., in which he writes, 'Visual poetry, sound poetry, theoretical text, encyclopedic information, fiction, lies, games, all are possible paths within the Interpoem.' (np).

12. Ibid.

13. *Mesostics for Dick Higgins* uses HTML techniques (HTML content is refreshed within a series of framesets) to animate a series of mesostic poems, a form whose development is attributed to Jackson Mac Low (diastics) and John Cage. Cage describes the technical characteristics of mesostic form in *I-VI*: 'Like acrostics, mesostics are written in the conventional way horizontally, but at the same time they follow a vertical rule, down the middle not down the edge as in an acrostic, a string which spells a word or name.' (64). Like aND's poem for Higgins, Cage's mesostics were ceremonial works, often celebrating fellow artists.

14. As explained in *Dictionary of the Avant-Gardes*, Higgins' book takes form as a collage, with four vertical columns across every two-page spread: 'One column continuously reprints critical essays, a second column poetry, a third theatrical scenarios (including *Saint Joan at Beaurevoir*), a fourth drawings' (286).

15. mIEKAL aND, 'Mesostics for Dick Higgins', english.umn.edu/joglars/mesostics/index.html.

16. For example, words used (sequentially) to fulfill the needs of the first line are: boreDom, Dinner, Drop, graDually, Difference, useD, galvanizeD, Do, freeDom, subsiDized, closeD, Doing, passeD, Degrees, closeD, heaD and Drop. The browser title bar is labelled 'the first 17 mesostics', and since upon close investigation 17 words are in each line's loop, we can presume aND composed that number of poems, isolated each to the proper line, and then automated them as described.

17. On a PC/Windows machine, this is done by pressing the 'Print Screen' button on the keyboard and then pasting the captured image into a graphics program. Using Mac OS X, pressing 'Command-Shift-3' automatically saves a file of the screen on the desktop.

18. aND, 'Mesostics for Dick Higgins.' April 23, 2009.

19. C. Funkhouser, *Prehistoric Digital Poetry: An Archaeology of Forms, 1959–1995*, (University of Alabama Press, 2007), 79.

20. Verbal information on the surface of the poem is drawn from a series of subdirectories (e.g. 'd/dinner.html'). Pasting a specific subdirectory address into the browser, each line will appear on its own. Doing so reveals that each word appears in the same order every time the piece is started, and that a small group of words is being recycled.

21. In *Seedsigns for Philadelpho*, the finite elements are manually arranged this piece is largely a digitally-processed analog-based pixilation. In order to accomplish these effects in the past, seeds would be drawn, or photographed, and sequenced as individual blocks; mesostic poems in *Mesostics for Dick Higgins* could have been rendered and sequenced in isolation, and soundtracks to both pieces could be provided.

22. In fact, *Mesostics for Dick Higgins* (Xerox Sutra Editions, 1998) can be read and downloaded via www.scribd.com/doc/40201627/Mesostics-for-Dick-Higgins-by-mIEKAL-aND.

23. collection.eliterature.org/1/works/andrews__stir_fry_texts.html.

24. www.vispo.com/arteroids/index.htm.

25. Before the WWW emerged, Jim Andrews' poetic practice included creating a series of static images, using graphics software programs such as CorelDraw, which appeared in a magazine he was then publishing titled *And Yet* (1990). These poems are now more likely to be viewed critically as 'visual poetry' than digital poetry, since their mode of presentation does not require a computer screen. Nonetheless, observing that software played an important role in composition, they were posited as 'Static Works' and 'Poems Fixed in Space' featuring 'Dispersal of Language' in *Prehistoric Digital Poetry*. Examples of Andrews' output from this period can be found on his extraordinary website, *Vispo* (www.vispo.com).

26. The most recent iteration of *Stir Fry Texts*, available at vispo.com/StirFryTexts, includes seven poems.

27. In at least two sections of *Stir Fry Texts*, this sense of voice can be taken literally, as each of the layers is composed by an

identified individual. To read Andrews' description of the structural
design of *Stir Fry Texts*, see 'Architecture of the Literary' (www.
vispo.com/StirFryTexts/architectureandtheliterary.html).

28. 'Log' is loosely framed within the context of an email exchange
     between Lennon and Andrews.

29. Jim Andrews, 'Stir Fry Texts,' collection.eliterature.org/1/works/
     andrews__stir_fry_texts/index.html.

30. Ibid.

31. Queneau's work was originally published as a book (1961)
     and has been digitally adapted by Tibor Papp in *Alire* 1
     (1989) and on the WWW via www.bevrowe.info/Queneau/
     QueneauHome_v2.html and www.growndodo.com/wordplay/
     oulipo/10%5e14sonnets.html (Feb 10, 2011). Christophe
     Petchanatz's 'Cut Up' in *Alire* 6 (1992) is another precursor to
     Andrews' work.

32. Jim Andrews, 'Stir Frys and Cut Ups', collection.eliterature.org/1/
     works/andrews__stir_fry_texts/text.html.

33. 'Spastext' is the only piece in *Stir Fry Texts* including multiple
     versions. A Chinese translation, which is beautiful to look at, was
     prepared by Dr. Shuen-shing Lee (2001).

34. 'Correspondence' incorporates Internet postings relating to the
     practice of recombinant textuality written by Andrews, Memmott,
     Lee Worden and Mary Phillips. In 'Divine Mind Fragment Theater',
     quotations by Jerome McGann (from *Social Values and Poetic
     Acts*), Joseph Weizenbaum (*Computer Power and Human
     Reason*) and Leo Marx (from 'The Struggle Over Thoreau') are not
     thematically related.

35. Andrews, 'Stir Fry Texts'.

36. Of note here are Marko Niemi's 'Concrete Stir Fry Poems',
     published by Andrews on Vispo (www.vispo.com/StirFryTexts/
     marko/index.htm). In this project, Niemi creates five interactive
     stages, in which the appearance of individual letters (black block
     letters on white background), or fragments of letters, is activated
     by mouseover. In most of the pieces ('accordionist' being the
     exception), at first the letters appear in linear order to spell out
     a word thematically united with a given title; depending on
     where the mouseover occurs, the visual/graphical arrangement
     may then devolve so as to present a combination of geometric
     fragments of the letters.

37. Andrews, 'Stir Frys and Cut Ups'.

38. Related techniques are used in Morrissey's *The Jew's Daughter*,

in which links are made (within a shifting block of prose) by
passing the mouse over hyperlinks programmed in Flash. Other
examples of digital poems that involve clicking on links to re-write
text atop a given visual scenario include Millie Niss' 'Sundays
in the Park' (see *Oulipoems*), geniwaite's *semtexts* (both made
with Flash) and Kenneth Goldsmith's *Soliloquy* (made with HTML/
DHTML), all of which are included on the *ELC* Vol. 1.

39.   *Arteroids* is not the only artistic rendition of *Asteroids*. *Calderoids*
(2005), produced by the Toronto collective Prize Budget for Boys,
uses the *Asteroids* motif, but instead of shooting/destroying
words takes aim at virtual models of Alexander Calder mobiles
(pbfb.ca/calderoids/). The Prize Budget for Boys website also
includes another repurposed art game, modelled after Pac-Man,
titled *Pac-Mondrian* (2004).

40.   Jim Andrews, 'Games, Po, Art, Play, & Arteroids 2.03', www.
vispo.com/arteroids/onarteroids.htm. Andrews uses Adobe
Director, Sound Forge (for audio) and Photoshop (for some
graphics) to construct his adaptation. In Andrews' rendering
of the interface and features of the game, some elements of
the original are absent, and others are added. Nick Montfort
encapsulates the major changes between the original and new
version in his essay 'Literary Games': 'large asteroids no longer
break into medium-sized ones which break into smaller ones, all
the while retaining their lethal power' and 'the absence of the
occasional flying saucer'. (np).

41.   A complete breakdown of the differences between the two
modes, and comments by Andrews about the significance of
the differences, is presented in the 'How to Play' section of
Arteroids.

42.   Jim Andrews, email, May 2009. The formula for calculating
scores is not disclosed anywhere in the game, but when asked,
Andrews writes: 'The number of texts you have to shoot to finish
a level is a function of the level. The higher the level, the more
texts you have to shoot to finish. The higher the level of play,
the higher your score can be. The score in *Arteroids* is mainly a
function of your speed and accuracy, however. The more accurate
you are – i.e. the fewer times you miss with your shots – the
higher the score. Also, the faster you complete play, the higher
the score.'

43.   Jim Andrews, 'Arteroids', www.vispo.com/arteroids/index.htm.

44.   Andrews explains in a Sept. 2009 email that he attaches sounds
of his voice – 'randomly pitch-shifted to increase the variety
of sound while not increasing the amount of downloaded

information' – to each of the keypad keys used in the negotiation, as well as to each of the 'explosions' that happen. As such, the keyboard is a percussive instrument in, and adds a sound poetry dimension to, the work. Without this noisy (sensory) aspect, which distracts users accustomed to silent reading, the association with the arcade game would be lost.

45. This device is accessed via the 'edit' button on the index.

46. Other default id-entities are 'poetry' (Andrews), 'Persephone' (McPhee), '/script' (Thorington).

47. Andrews, 'Arteroids'. These phrases appear in the 'How to Play' section.

48. Users have the option to make both the same, and once this choice has been selected, the original default text is no longer there.

49. 'Mortal' and 'deathless' settings are established here.

50. Andrews, 'Arteroids'.

51. Noted by Andrews in a Sept. 2009 email is the fact that the dynamics in 'Word for Wierdos' enable an unusual bi-partite literary form: There are five texts: the outer green, the inner green, the outer blue, the inner blue and the id-entity. Each line of the outer green defines a text that appears on-stage in green. The corresponding line of the inner green text is what that outer green line explodes into when the outer green text is shot. Similarly, each line of the outer blue text defines a text that appears on-stage, and the corresponding line in the inner blue text is what the outer blue text explodes into when its line is shot.

52. Those readers curious to know more about this type of work should see Nick Montfort's essay 'Literary Games', which introduces several other examples.

53. In his essay 'Arteroids, Poetry, and the Flaw', Andrews acknowledges that there is tension between poetry and game. He makes the significant point that 'poetry is not a game somebody wins', and thus *Arteroids* equally explores the ways in which poetry/art is not a game-like thing (np).

54. Neil Hennessey, 'Basho's Frogger', collection.eliterature.org/2/works/hennessy_frogger_jabber.html.

55. See www.bopsecrets.org/gateway/passages/basho-frog.htm for a series of translations of the poem. This particular haiku was previously distilled by the Concrete poet Dom Sylvester Houedard to 'Frog-Pond-Plop' (1965).

56. Andrews, 'Arteroids'.

57. Note: sound bytes used in *Arteroids* have also been utilized by Andrews in a recent project titled 'Jig-Sound' (vispo.com/temp/jigsoundArteroids2b.htm). For the sound component in this interactive tool, Andrews has selected three dozen samples from *Arteroids* to use as a basis for looping and sequencing the sounds he provides.

58. Andrews, 'Arteroids'.

59. collection.eliterature.org/1/works/cayley__wotclock.html.

60. Cayley has, on occasion, characterized his digital works as 'machine modulated', especially earlier in his career; his practice of 'transliteral morphing' is discussed both in *Prehistoric Digital Poetry*, and in Hayles' *Electronic Literature*.

61. At present, Cayley, with Daniel C. Howe, is developing *The Reader's Project* (thereadersproject.org/), which holds great promise as it 'explores and visualizes existing and alternative *vectors* of reading, vectors that are motivated by the properties and methods of language and language art' (Readers). Cayley has also created 3D textual works in Brown University's CAVE, an immersive virtual reality environment. *Torus*, a work initiated in 2004, is still in progress. Another CAVE work, *Lens* (homepage.mac.com/shadoof/lens/lens.html), was repurposed for the WWW in 2006. In the online version, words are presented in different layers, whose colours do not always contrast well. A 3D effect is simulated but not fully experienced. Presentation of language in this form projects a commentary on legibility and the necessity of developing reading strategies when textual surfaces are inverted, the impact of which is greater when experienced in 3D. The poem becomes about one's orientation within the poem, and the different layers and meanings contained within the given information. He is author of several important critical essays. Treatises about coding as writing ('Coding as Practice') and the nomenclature of 'electronic literature' ('Weapons of the Deconstructive Masses') were published in 2007; the latter argues for 'writing digital media' as an alternative designation to describe the naming for what digital writers do.

62. Douglas Cape and Giles Perring, 'What We Will', www.z360.com/what/.

63. A visual timeline featuring two full panoramas taken from the top of St Paul's Cathedral (one during the day, one at night), superimposed with Roman numerals 1–24 is presented at the top of the screen; clicking on it brings new information (from a

different time of day) to the screen, e.g. high-resolution, drag-able panoramic photographic scenes appear, as does a descriptive text in a pop-up box, and a looping soundtrack.

64. J. Cayley, 'P=R=O=G=R=a=M=M=a=T=O=L=O=G=Y: Wotclock', programmatology.shadoof.net/index.php?p=works/ wotclock/wotclock.html. In a 2010 interview Cayley told me the text(s) that appear in *Wotclock* are coded (extracted, put together) in QuickTime but were pre-generated using a Markov model that constructs bigrams (two word phrases) from the text that can be heard in the spoken dialogues he composed for the corresponding twenty-four scenes (hours) of *What We Will*.

65. Cayley's research and literary activity have included many translations of classical Buddhist and Taoist poetry (e.g. Li Bai, Du Fu, Qian Qi), as well as contemporary 'underground' Chinese poets (e.g. Yang Lian). In my view, Cayley's knowledge of the motifs of contemplative (Eastern) expression directly influences his work.

66. As Cayley explains in 'P=R=O=G=R=a=M=M=a=T= O=L=O=G=Y: Wotclock', 'a letter representing any part of a number for the time of day is shown in red. If no letter within a word is red, the implied number is zero. The first two words in the centre of the clock give the hour; the second two give the minute.' Cayley notes that the clock's algorithms 'generate the phrase-texts in a Markov chain derived from underlying composed dialogue fragments for each hour-scene. The chain-making is then, where necessary, overridden by the encoded, mesostic time-telling requirements.' (Ibid.).

67. John Cayley, 'Wotclock', collection.eliterature.org/1/works/ cayley__wotclock.html.

68. Ibid.

69. Ibid.

70. Further, since Cayley appears in the scene (holding an envelope in the coloured image), readers may make an intuitive connection: that he, beyond being the author, is a character in the drama.

71. Cayley, 'Wotclock'. May 21, 2008.

72. Here I am referring to works such as Glazier's *IO Sono at Swoons* (epc.buffalo.edu/authors/glazier/java/iowa/), which will play endlessly but more or less appears the same way throughout. Tisselli's *Degenerative* website is also ever-evolving, slowly becoming corrupted ('each time the page is visited, one of its characters is either destroyed or replaced'), but destroyed figures

never reappear and the site appears in the same unreadable state forever (www.motorhueso.net/degenerative/). Andrews' *dbCinema*, which visually changes over time, eventually runs in a loop if users do not make manual adjustments.

73. collection.eliterature.org/1/works/larsen__carving_in_possibilities. html.

74. This self-assessment appeared on Larsen's now non-existent WWW project *Datafeeds*, formerly housed at www.deenalarsen. net/datafeed/index.htm. In 1993, Eastgate Systems published Larsen's hypertext poem *Marble Springs* on floppy diskette, and also published her hypertext stories 'Century Cross' (*The Eastgate Quarterly Review of Hypertext*, Vol. 2 No. 2) and *Samplers* (1998).

75. Deena Larsen, 'Deena Larsen's Addicts Attic: Hypertext/New Media/Electronic Lit/Possibilities', www.deenalarsen.net. Her website is rich with information, links to artworks, and also includes a section titled 'Fundamentals: Rhetorical Devices for Electronic Literature', a valuable introductory handbook for anyone interested in electronic literature. Larsen's unpublished titles are archived at Virginia Tech's Center for Digital Discourse.

76. Deena Larsen, 'Carving in Possibilities', collection.eliterature. org/1/works/larsen__carving_in_possibilities.html.

77. Ibid.

78. Ibid.

79. Ibid.

80. Ibid.

81. Ibid.

82. Ibid., April 28, 2009.

83. Larsen has employed mapping techniques in her online works, such as 'The Pines at Walden Pond' (see www.cddc.vt.edu/host/deena/pines/).

84. Larsen, 'Carving in Possibilities'.

85. I.e., in some viewings, more than seventy links (texts) appeared before the narrative ended; in others, about thirty. Moving the mouse over the lower part of the picture seems to make the evolution of image occur more quickly. In response to my inquiry about this effect, in an April 2009 email Larsen writes: 'When I designed this, the pace was around 70 frames, but newer versions of the Flash player may read the .swf file differently from earlier editions. Some of the buttons' hit areas were smaller than others', she adds, 'so the mouse could be hitting texts more

rapidly. This is part of the interaction, so that the speed of reading differs through each reading.'

86. Larsen, 'Carving in Possibilities'.

87. Ibid.

88. collection.eliterature.org/1/works/rosenberg__diagrams_6_4_ and_10.html.

89. *Mesolist*, a program he developed in the early 1980s, was used by John Cage to compose mesostic poems (1984). His groundbreaking interactive diagram poems *Intergrams* were published as the vol. 1.1 of the *Eastgate Quarterly Review of Hypertext* (1993); two others, *The Barrier Frames and Diffractions Through*, appeared as vol. 2.3 of the same publication (1996).

90. Jim Rosenberg, email, June 2008. Rosenberg's Diagram Poems were initially static, such as Diagram Series 3 (see www.well. com/user/jer/d3/diags3.html).

91. Judy Malloy, Anna Couey and Douglas Cohen, 'A Conversation with Jim Rosenberg', www.well.com/~couey/interactive/jim.html. In this interview, Rosenberg explains that his process involves (1) maintaining 'reservoirs' (precompositional groupings of phrases), which are (2) permuted via chance and other operations into subsequent reservoirs. He employs (3) an 'edit phase' in which many phrases are further metamorphosed, rearranged and reassembled. After completing this series of textual-technical phases, Rosenberg subjectively chooses the phrases that appear in the poems.

92. These concepts are discussed at length in essays 'Navigating Nowhere/Hypertext Infraware' (1994), 'The Structure of Hypertext Activity' (1996), 'The Interactive Diagram Sentence: Hypertext as a Medium of Thought' (1996), 'And *And*: Conjunctive Hypertext and the Structure Acteme Juncture' (2001), and other writings linked to his website (www.well.com/user/jer/poetics.html).

93. Hélène Perrin and Arnaud Regnauld, 'Jim Rosenberg's Diagram Poems Series #3: A Few Preliminary Notes on Translation Issues', in *E-Poetry 2007: An International Digital Poetry Festival* (Paris, France: Proceedings of E-POETRY 2007).

94. Ibid.

95. Jim Rosenberg, 'The Interactive Diagram Sentence: Hypertext as a Medium of Thought', *Visible Language* 30.2, New Media Poetry: Poetic Innovation and New Technologies, 116.

96. Jim Rosenberg, 'Diagram Series 6: 6.4 and 6.10,' collection. eliterature.org/1/works/rosenberg__diagrams_6_4_and_10.html.

97. Perrin and Regnauld, 'Jim Rosenberg's Diagram Poems Series #3: A Few Preliminary Notes on Translation Issues,' 1.

98. Rosenberg, e-mail, August 2008

99. Perrin and Regnauld, 'Jim Rosenberg's Diagram Poems Series #3: A Few Preliminary Notes on Translation Issues', 14.

100. In *Geeks Bearing Gifts*, Nelson celebrates Squeak as 'a complete programming environment for everything' (102).

101. *Diagrams Series 5* and *Diagrams Series 6* are available via www.well.com/user/jer/inter_works.html.

102. Rosenberg, 'Diagram Series 6: 6.4 and 6.10'.

103. Ibid.

104. Perrin and Regnauld, 'Jim Rosenberg's Diagram Poems Series #3: A Few Preliminary Notes on Translation Issues', 5.

105. Rosenberg, 'Diagram Series 6: 6.4 and 6.10'.

106. Ibid.

107. A point that needs to be stressed is that Rosenberg's diagram syntax can contain feedback loops. Ordinary syntax does not allow feedback loops, but diagram syntax makes them very easy to draw – even if their function as such is undefined. The 'outer' arrangement in Diag. 3.12 is obviously a loop. While not as overt, Diag. 3.13 also contains a loop: the two rectangles inside the larger rectangle are joined by relation and the verb is the complex joining them and activating their relation. This construct, writes Rosenberg in an August 2008 e-mail,

    represents a type of feedback loop in which there is a relation between a part and the whole in which it participates. (The rectangle is a 'scope:' when it participates in the syntax, the item at that 'node' is the contents of the rectangle.) So ...you have two complexes joined by a relation in which the verb is the whole relation – the verb is the relationship itself. This is an example of an 'internal relation', which is a particular form of feedback loop, one which I have been fascinated with over the years and have used many times.

    What role these feedback mechanisms play here or anywhere in Rosenberg's Diagram poems is uncertain. How many times can a verbal complex feed back into or through itself, and to what end? This type of mechanism seems to promote or encourage multiple, variant readings of the same material in order to glean different results. Presumably, however, while many interpretations will be possible, depending on order of access, an infinite number of meanings cannot be inferred through such a limited amount of verbal material, particularly since all of it is intended for simultaneous absorption. At a certain point, the complexity of Rosenberg's conceptual operations begins to interfere with practical reception; deep readers will need to determine at what point activating such feedback as readers inhibits their forward movement through the poem.

108. Rosenberg, 'Diagram Series 6: 6.4 and 6.10'.

109. Ibid.

110. Perrin and Regnauld, 'Jim Rosenberg's Diagram Poems Series #3: A Few Preliminary Notes on Translation Issues', 6.

111. Rosenberg, 'Diagram Series 6: 6.4 and 6.10'.

112. Ibid.

113. Ibid.

114. Ibid.

115. Rosenberg, 'The Interactive Diagram Sentence: Hypertext as a Medium of Thought', 112.

116. Ibid., 113.

117. Perrin and Regnauld, 'Jim Rosenberg's Diagram Poems Series #3: A Few Preliminary Notes on Translation Issues', 3.

118. Ibid., 12.

119. Rosenberg, 'Diagram Series 6: 6.4 and 6.10'.

120. In a August 2008 email, Rosenberg responds to this observation:

    That brings up a very tricky issue, which is what happens when I give a reading of a diagram Is the structure 'destroyed'? In a way, alas, the answer is yes. I suppose I should speak about this when I give readings. I think of a reading from a work like Diagrams Series 6 as pictures from a catalogue: it shows you some things that are there in the work, but isn't really 'the work'. And of course: I'm not reciting the diagram symbols – I haven't figured out a way to do that that isn't hokey.

121. www.alansondheim.org/.

122. collection.eliterature.org/1/works/strasser_sondheim__tao.html

123. collection.eliterature.org/1/works/strasser_sondheim__dawn.html.

124. As noted in *Prehistoric Digital Poetry*, codework is a term coined and developed by Sondheim in his introduction to a series of articles published in the *American Book Review* in 2001.

125. Materials collected in *Internet Text* are produced in a range of forms and correspondent filetypes, including writing (.txt,

.pdf, .htm), images (.jpg, .png), sound (.mp3), video (.mov, .avi, .mp4) and hypertext (.htm). A hundred and ninety-one files from *Internet Text* were compiled and included as part of the *ELC* Vol. 1.

126. *Internet Text* contains 18 hypertext works, including 'economies of the imaginary', a hypertext poem. Many of these pieces (e.g. 'Jennifer,' 'It hurts you,' 'how could you') are minimally interactive but use viewer response to transform screen content.

127. Alan Sondheim, 'On the Organization and Disorganization of Files', www.alansondheim.org/00README1st.TXT. Sondheim explains organization that occurs happens via the filenames, 'in the form of the name itself, what follows in parallel'.

128. Ibid. Sondheim explains his approach to presenting the copious materials, 'The world is such a skein of discovery, discomfiture, self-organizing systemics among perceptions and perceptual algorithms. Here, one employs the same; everything is available, the extensions guiding format and sense, the contents interrelated across extensions through parallel names.'   .

129. Ibid.

130. Ibid.

131. Ibid. Nikuko and Jennifer are the names of two avatars that prominently appear in the 'long wave' segments. See 'Avatars' for an introduction to their personas.

132. Alan Sondheim, 'Blood', www.alansondheim.org/Blood.txt.

133. Ibid.

134. Both works are included in the *ELC* Vol. 1. The *ELC* title page credits Sondheim and Strasser as co-authors. The title page of 'Dawn' refers to the work as 'Alan Sondheim's Dawn', crediting Sondheim for text, photography, sound and Strasser for Flash composition.

135. Alan Sondheim and Reiner Strasser, 'Dawn', collection.eliterature.org/1/works/strasser_sondheim__dawn.html.

136. Ibid.

137. Ibid. The authors' note describes the work as combining, 'aspects of love, death and nature'.

138. Ibid.

139. Ibid.

140. Alan Sondheim and Reiner Strasser, 'Tao', collection.eliterature.org/1/works/strasser_sondheim__tao.html.

141. Ibid.

142. Sondheim, 'On the Organization and Disorganization of Files.'

143. Second Life is built on servers and servers' farms, different ones for different 'islands,' which are cached on your own machine; it is computing intensive, and takes up a lot of resources.

144. The inscribed architecture of this virtual world is sometimes surreal, but often emulates structures from 'real life'; it never rains there, but automatically generated, programmed 'winds' can blow things around, and users can set up grids that respond to it.

145. Sondheim reported in our 2010 interview that it did not take a lot of time for him to prepare interesting materials (visual, 3D, interactive) once he had a sense of what he was doing, what vocabulary he had to work with; individual scenarios were not time consuming to make.

146. Documentation of dozens of Sondheim's experiments in Second Life archived on his YouTube page, www.youtube.com/user/asondheim, were unfortunately removed in 2011. Examples are available through *Internet Text*, and www.youtube.com/user/BeamMe09. He also produces similar works for Odyssey art: see odysseyart.ning.com.

147. Sandy Baldwin, 'Against Digital Poetics', *E-Poetry 2009: An International Digital Poetry Festival* (Barcelona, Spain 2009).

148. Ibid.

149. Alan Sondheim, Personal interview, Brooklyn, NY, Feb. 2010.

150. Ibid. For example, when showing an example of some 'eerie trees' he created, he comments, 'I don't know what words I could use to describe them.'

151. Ibid.

152. Alan Sondheim and Sandy Baldwin, *Second Life Performance*, 2009. Digital Poetry performance: E-Poetry 2009: An International Digital Poetry Festival.

153. In the concluding chapter of *Prehistoric Digital Poetry*, Baldwin's 'New Word Order' (2003), a modification of the computer game *Half-Life*, is discussed as a representation of the movement towards a cybertextual mode in digital poetry (245). Documentation of 'New Word Order' is published in the *ELC* Vol. 2.

154. Sondheim and Baldwin, *Second Life Performance*.

155. Ibid. Further, in our 2010 interview, Sondheim reported he is 'not interested in generative writing unless it is something I can modify – wouldn't want to have a program that was just generating text'. Asked about his role in the presentation of texts,

Baldwin explained in his 2008 interview how he 'greps' and searches Sondheim's work, orders it in different ways, and that some is Google-sculpted, going along with themes in Sondheim's work.

156. Ibid.

157. Alan Sondheim, *Self Induction,* 2008.

158. Partial documentation of this example may be viewed via harp. njit.edu/~newrev/3.0/Sondheim_shudder.mp4.

159. Alan Sondheim, 'Shudder', (2010).

160. Sondheim and Baldwin, *Second Life Performance*.

161. Sondheim. Discussing challenges he has encountered using text in Second Life in his 2008 interview, Baldwin reports:

> I have about 100 different commands that can trigger combinations of text and gestures. They're hard to keep track of, but as I added commands I got more of a sense of the theatrical; our earlier experiments don't have that. I evoke them, but collaborating in Second Life you're not seeing things from the same angles, and an audience is in third position – you have a highly intimate performance, which is codified, and yet nothing ever touches. You have this totally technical production, and it is cumbersome because I can't see where he is, and the space is so cluttered. I have all these spheres around me. Sometimes there are other people there. This isn't a private space, which is great. You don't know where people are coming from.

162. Sandy Baldwin, Personal interview, Oct. 2008.

163. Ibid.

164. Sondheim.

165. Sondheim noted in our interview that sometimes arenas where people 'sit' are set up, so one can have audience in fixed chairs, which helps with dance and formalized theatrical pieces, but recalled an occasion where seats were set up in mid-air, and an audience was moved as a whole through a remake of the space of *Metropolis*.

166. Caterina Davinio, another digital poet artist who has engineered works in Second Life, in 2009/10 created an installation on the network to celebrate the centenary of Italian Futurism and the 40th Anniversary of the first Moon landing, calling it the 'First Poetry Space Shuttle Landing on Second Life'. In this work, the poets take on the form of spheres and colourful prisms which contain and project poem files in the virtual space. The guest book of the shuttle 'Welcome on Board' contains more then 300

poems by 230 authors from various continents, countries and planets. See www.youtube.com/watch?v=E3zMgGMiUuw for documentation of this project.

167. In 2010 Sondheim delivered a terabyte of materials, including 30,000 still images, to an archive at New York University.

Notes to Chapter 4

1. www.to-touch.com/.

2. lossofgrasp.com/.

3. Serge Bouchardon, Kevin Carpentier and Stéphanie Spenlé, 'Touch', www.to-touch.com. *Touch* does not include directions per se, but offers a link to a detailed description, from which viewers can easily derive instructions.

4. At all times a 'Back' button may be used to return to the main interface.

5. Bouchardon, Carpentier and Stéphanie Spenlé, 'Touch.'

6. Ibid.

7. For example, as viewers move the mouse over these slots, the line shifts to form phrases such as, 'Do you touch me when you touch me' or 'Do I touch you when you touch me'.

8. Bouchardon, Carpentier and Spenlé, 'Touch.'

9. Ibid.

10. Ibid.

11. Ibid.

12. Ibid.

13. Excerpts from Aristotle's text featured in *Touch* are from Book IV, Section 8 and Book IV, part 7; the former passage speaks to the sense of touch being 'common to all animals whatsoever'.

14. Using headphones, viewers will sense that it circles her/his head.

15. Serge Bouchardon, 'Presentation: Touch' (2009), www.utc.fr/~bouchard/TOUCHER/docs/presentation-touch.pdf.

16. Unless they use the 'Back' button, users cannot further progress through the piece.

17. Bouchardon, Carpentier and Spenlé, 'Touch.'

18. Ibid.

19. The presentation is somewhat imperfect, however, because if someone's hardware (microphone) and software (Flash player) do not communicate properly, the viewer is left to watch snowflake patterns fall from top to bottom on the stage. In this research, I was unable to sync the two via to-touch.com, but was able to make it work via its appearance in *The New River* (Spring 2010); see: www.cddc.vt.edu/journals/newriver/10spring/thepieces/bouchardon-touch-newriver/index.html.

20. Bouchardon, Carpentier and Spenlé, 'Touch'.

21. To note another example, in a 2009 offline performance in Bergen, Norway, Simon Biggs effectively presented a vocal-response work, in which inaccuracies of the software's rendering of language on the screen were just as much a part of the show as the precision.

22. Bouchardon, Carpentier and Spenlé, 'Touch'.

23. Bouchardon, 'Presentation: Touch'.

24. Ibid.

25. Ibid.

26. Ibid.

27. Serge Bouchardon and Vincent Volckaert, 'Loss of Grasp', lossofgrasp.com/.

28. Upon activation, a welcome message from a female voice greets viewers and tells them to 'press the hash key', and congratulates them once they have.

29. This is noteworthy because few 'preloaders' for Flash productions contain sound; Bouchardon's is the only one I have seen. When a large Flash file is loading, authors commonly offer a visual representation of its progress.

30. Bouchardon and Volckaert, 'Loss of Grasp'.

31. Ibid.

32. Ibid.

33. Ibid.

34. Ibid.

35. Ibid.

36. Ibid.

37. Ibid.

38. Ibid.

39. Ibid.

40. Ibid.

41. Ibid.

42. Ibid.

43. Ibid.

44. Ibid.

45. Ibid.

46. Ibid.

47. Ibid.

48. Ibid.

49. Ibid.

50. Ibid.

51. Ibid.

52. Two examples of this type of work are Todd Winkler's *Light Around the Edges* (1997) and Max Kazemzadeh's *Target Audience* (2004).

53. The only other work I know of that contains such a webcam attribute, apart from augmented reality works, is Bouchardon's *Touch* (2009).

54. Bouchardon and Volckaert, 'Loss of Grasp'.

55. Ibid.

56. Formerly housed at etc.wharton.upenn.edu:8080/Etc3beta/Erika. jsp.

57. *Erika* had previous iterations in Python, C++ and Java platforms.

58. Jim Carpenter, 'Erika', etc.wharton.upenn.edu:8080/Etc3beta/ Erika.jsp.

59. Ibid.

60. Many examples of *Erika*'s output exist. Carpenter's program was used by Stephen McLaughlin, Gregory Laynor and Vladimir Zykov to create poems published in *Issue 1*, a 3,785-page e-book (.pdf) purporting to feature works by more than 3000 poets. None of the poets to whom works were attributed had actually written the poems, however, and the publication provoked a scandal and much dialogue about the project across the WWW. A four hundred page compilation of dialogue about *Issue 1*, collected from blogs and websites by Erika Staiti, was published as *Issue 2* (saidwhatwesaid.com/); another document compiling dialogue about the project that appeared on SUNY-Buffalo's POETICS listserv is accessible via the same URL. Those interested in learning more about how *Erika* works should see Mikael Hvidtfeldt Christensen's discussion of the program on Christensen's blog, *Syntopia*, blog.hvidtfeldts.net/index.php/ category/poetry/.

61. Carpenter, 'Erika'.

62. Detailed information and directions about how to use each of the categories were included below the table.

63. Carpenter, 'Erika'. In his notes on the work, Carpenter offered an explanation: *Erika*, which he refers to as Etc3, takes the input text and 'parses it to discover words it recognizes and can reasonably use to construct poetic utterances. It ignores words it doesn't recognize. If you asked it to write a poem about sub-prime mortgages, it would just ignore your topic, since neither of those words is in its semantic model – it would, however, write something on a topic of its choosing.'

64. Ibid. Information about each stanza type was provided. Different settings achieved different results. For example, the Generic Open Form 'breaks utterances into short lines, indenting another tab stop at each break'; Paratactic 'selects one syntactic structure from the grammar you select and then uses that for each of its lines.' A stanza type that is discussed but not available through the interface provided, 'Grammar test stanza', was also introduced.

65. Ibid. Carpenter defines the method of each grammar, and offers suggestions for use, in accompanying information (as above).

66. Ibid. Carpenter explains, 'A word pool is a 'bag of words' that directs Etc3's morphological decisions', noting that different kinds of word pools are available and that the user may select several at once: *Erika* 'will create a pool that is an amalgam of your selections.' Strictures of each type of word pool are provided, e.g. 'Topic antonyms' are 'a collection of words other than topic words but with approximate opposite semantic meanings'.

67. Ibid. Preferred tenses were present, past, future, present perfect, past perfect, future perfect; preferred subjects consisted of all pronouns; preferred objects were her, me, us, you, them, him and it; 'preferred tense' indicates the program 'will try to use the tense you select, but some TAG trees will override it'; for preferred subject and object, she 'will try to use the value you select'.

68. Ibid.

69. Ibid.

70. Perhaps coincidentally, a blind poet named Laura Redden Searing (b. 1839) was a contemporary of Dickinson's.

71. Carpenter, 'Erika'.

72. Ibid.

73. On basic mechanical levels, such as number of stanzas and lines, the program performed perfectly.

74. Viewers could add words into, or subtract them from, the program's database since the entire work was open source.

75. *Prehistoric Digital Poetry* thoroughly observes the historical development of this form, and builds associations between digital and print-based poetic inventions.

76. Carpenter, 'Erika'.

77. www.cddc.vt.edu/journals/newriver/08Fall/mapNewRiver/mapNewRiver/_MapOfAFutureWar.html.

78. angelaferraiolo.com/end_of_capitalism/capitalism.html.

79. Artist Biography: 'Angela Ferraiolo', www.brown.edu/Conference/Electronic_Literature_Organization/bios.php.

80. Angela Ferraiolo, 'Map of a Future War', The New River, www.cddc.vt.edu/journals/newriver/08Fall/mapNewRiver/mapNewRiver/_MapOfAFutureWar.html. This description appears on the author's homepage, into which *Map of a Future War* is embedded; see: www.angelaferraiolo.com/map_of_a_future_war/map.html.

81. *In absentia* (2008), by J. R. Carpenter, is another hypermedia literary work that more literally uses maps and mapping technologies to propel narrative. Addressing 'issues of gentrification and its erasures in the Mile End neighbourhood of Montreal', Carpenter prosaically fashions a collaborative narrative (with HTML, JavaScript, and CSS) utilizing images provided by Google Earth as a hypertextual mechanism.

82. Ferraiolo, 'Map of a Future War'.

83. Ibid.

84. Ibid.

85. On many screens variances in scale of font, which dictates what comes first to a viewer's eyes, are not detectable. In them, unlike the opening screen, the size of text is more or less equal.

86. Here I am referring to writers such as Poe, Hemingway, Kerouac, Bukowski and the like.

87. Ferraiolo, 'Map of a Future War'.

88. At an explicit and detailed moment in the text, the author graphically refers to the Government form 230i, which is an application for immigrant visa and alien registration.

89. Angela Ferraiolo, email, July 2010. Describing the compositional arrangement, Ferraiolo confirms this assessment is accurate:

Map was built by layering visuals over a big grid, about twelve times the size of the computer screen. In other words, there is one very large base grid and as the reader selects a hexagon, the computer acts as viewport for a giant canvas that slides around in all directions. Text and graphics are added and removed as the canvas comes in and out of view. Some of the graphic elements of the canvas are fixed, others change along with the text. There are twelve grids and each of the grids has four presentational modes. There are about two hundred text blocks, maybe five hundred graphic elements.

90. Ferraiolo, 'Map of a Future War'.

91. Ibid.

92. Ibid.

93. Ibid.

94. Ibid.

95. According to Ferraiolo (email, July 2010):

It's true that these changes give the story a feeling of progression, but there's no equation leading to an end point. It's not so much that I was trying to have the story continue on indefinitely as that I wanted to write something that we could call spatial, a story that takes its narrative model not from character arcs or sequential plot but from Number 1A or a Calder mobile. Ideally, *Map* [*of a Future War*] would be experienced like a sculpture, as a constellation of simultaneous fictional moments. I did not really come up with this idea on my own, though. If you look at some of Virginia Woolf's short stories, you'll see that she is sort of going after this same effect. In *Map* I exaggerate the idea of the spatial by arresting time to a more radical degree and shifting the narrative point of view more rapidly.

96. Ferraiolo, 'Map of a Future War'.

97. Angela Ferraiolo, 'The End of Capitalism', angelaferraiolo.com/end_of_capitalism/capitalism.html.

98. Ibid.

99. Ibid.

100. On the introductory page, Ferraiolo also links to a series of eighteen screenshots made from the work, all but two of which are devoid of text. Texts that do appear target thematic concepts found in the piece: 'Dollar is a/symbol' and 'an American/tradition'.

101. Ferraiolo, 'The End of Capitalism'.

102. Ibid.

103. Within them, the entire stage is active; viewers can click on the screen at any time to proceed further. If they do not, the text remains for a short period and eventually proceeds on its own to a new segment.

104. Ferraiolo, 'The End of Capitalism'.

105. Ibid.

106. Ibid.

107. Ibid.

108. In the third stated possibility, the poem would read, SPITEFUL/A/ CYCLE/NOBODY READS/ANYMORE/LITTLE PROTECTION/ REVEALED/POOR TIMING.

109. Ferraiolo, 'The End of Capitalism'.

110. Ibid.

111. After viewing this title a few times, sometimes clicking through, sometimes not, I began to interact more with it when I began to encounter long segments I had already seen. I did not discover any sort of new content that arose from participating in the experience of the text in this manner.

112. Ferraiolo addresses the concerns and contents of this paragraph's comments in a July 2010 email, writing:

   [The End of] *Capitalism* tries to move outside time by confining itself to one long introspective look on the part of an unnamed narrator. (In the version posted, there are about sixty text spaces and seventy video rotoscopes. The shortest rotoscope is three frames and the longest is a little over two hundred and fifty frames.) It is not so much that an active viewer will have come across more content or even different content, but that an active viewer will be producing his or her own strategy of progression, reading the piece in a way that is their own, that privileges a personal intuition of the narrative. Both pieces [*Map of a Future War* and *The End of Capitalism*] were sort of explorations in that I was looking for a kind of interactivity that would reinforce narrative, an interactivity that's different from the interaction found in games.

113. Lev Manovich, 'Soft Cinema: Ambient Narrative', www.softcinema.net.

114. maryflanagan.com/house.

115. maryflanagan.com and www.tiltfactor.org/.

116. By this I mean viewers do not experience direct address, but are rather being let into something external.

117. Based on the spatial vector, the colour of the box changes as does the view of the words.

118. Mary Flanagan, '[theHouse]', collection.eliterature.org/1/works/flanagan__thehouse.html.

119. Ibid.

120. Flanagan mentions, though does not explain, that [theHouse] is 'regulated by the computational process of the sine wave' in her introduction to the work.

121. Flanagan, '[theHouse]'.

122. Ibid. Phrases including 'hand' appear in other contexts as well, e.g. 'with nothing in my hands/filled with questions'.

123. Ibid.

124. Ibid.

125. Ibid.

126. Ibid. N.B. [theHouse] does not include sound.

127. Ibid.

128. 'Processing', processing.org/. According to the Processing website, the program is 'used by students, artists, designers, researchers and hobbyists for learning, prototyping and production. It is created to teach fundamentals of computer programming within a visual context and to serve as a software sketchbook and professional production tool.... [it] is an alternative to proprietary software tools in the same domain.'

129. Other digital poems sharing aesthetic similarities to this work include Aya Karpinska's *arrival of the beeBox* (moveable, layered, text in motion) and Jason Nelson's 'cube creation', *Birds Still Warm from Flying* (manipulation of various sorts of layered information). In Karpinska's work phrases appear on wheels in three visual planes, which can be juxtaposed in many different ways and at variable speeds. Nelson's piece, rich with media, e.g. video and sound, is less nebulous, built on the model of the Rubik's cube, and viewers have control of direction and rate of reading the text.

130. www.hotkey.net.au/~netwurker/xor/xor.html.

131. Sandy Baldwin, 'Against Digital Poetics', *E-Poetry 2009: An International Digital Poetry Festival* (Barcelona, Spain 2009). Baldwin continued his commentary by observing that these works, while included in discussions of electronic poetry under the rubric of 'codework', are never seen as the leading edge and proclaimed his opinion that they 'are the best and perhaps only examples of electronic poetry'. These comments do not appear as such in the version of the essay later published in the *Electronic Book Review*.

132. A range of writers, from different poetic disciplines, have experimented with word and symbol play. Beyond e e cummings, who is perhaps the best known and most popular, poets as diverse as Norman Pritchard, Joan Retallack, Bill Bissett and Gabrielle Welford have published works featuring unconventional hybridized approaches to composition.

133. MUD (Multi-User Dungeon) was a pre-WWW real-time, text-based, multiplayer online environment.

134. Mary Anne (Mez) Breeze, 'Twitter', twitter.com/netwurker. The author herself, in a 2009 Twitter post, contemplates 'how repeated 'mezangelleisms' forge themselves a standardized 'codetic' (code>poesis combo) lexicon (think:'RIP' '[s]kin' etc).'

135. 'Mezangelle', en.wikipedia.org/wiki/Mezangelle. Mez's work tends to fuse biological/physical and online living; recurring themes she has addressed since 1996 include 'gamer dynamics, social engineering, questioning conceptions of print-based and electronic literature, ASCII art, play theory, teledildonics, viral imagery and examinations of postmodern, feminist, neural net, social change and technofetistic theories'. Her *free.form] [ulation][* website packages links to her projects, biographic and bibliographic information, past and present blogs, a refined portfolio, poetics statement and curriculum vitae in a single location. All of Mez's online works – Flash movies, code works, early hypertexts and other interactive and animated works – are connected to this webpage.

136. The text for this work (without media elements) first appeared on the Internet on the Rhizome listserver, February 2004; it was repurposed for the WWW later, and published online in *Hyperrhiz* 3 (2007). This work was also included in a virtual exhibition named dam(b)a({n}d{a})ged poetry. Its companion piece ('Wurk 2 in the W[n]e[t]b.Wurks_Series'), *Types.of.Und.Fineable.Ware[z]*, is less complicated; though more ornately using mezangelle, it contains less verbal information and fewer, less distorted images, giving readers fewer implications to consider. Mez has orchestrated a number of different types of interactive work on the WWW, including *Shutters of Defunct Meat (_][s][hut][ters][of d.funct meat_)*, in which imaginative use of roll-over images and text (approximately ten in all) has the effect of transforming the initial messages presented.

137. The word 'Java' appears symmetrically within at least one of the graphical images in Mez's horizontal scroll, most plausibly as some sort of acknowledgement to the technology which allows her to make such a unique composition. Perhaps more specific

connotations are intended, although I cannot venture to suppose what they are.

138. Dash, pipe, bracket and 'less than' symbols communicate visually in this section of the work.

139. Mary Anne (Mez) Breeze, 'Id_Xor-Cism', www.hotkey.net. au/~netwurker/xor/xor.html.

140. Ibid. A note on notations used for these passages: '/' indicates a separation or spacing between the linear verbal passages; boldface indicates pink text.

141. Ibid.

142. Projecting luxuria might be something like exhibiting to others hedonistic desires; whereas projectiling luxuria suggests and sounds more like direct imposition.

143. Breeze, 'Id_Xor-Cism'.

144. Ibid.

145. Ibid.

146. Ibid.

147. 'Simulcra for Amiga', www.mobygames.com/game/simulcra. *Simulcra* is described as follows on the MobyGames webpage: 'The *Simulcra* unit was thought to be the last best hope for peace. International disputes could be battled out in a simulated environment, in a world with no real guns or bombs. Unfortunately, the system malfunctioned and began generating real weapons and tanks to attack the outside world.' (Simulcra). Players of the game are 'charged with putting a stop to all of this'.

148. Breeze, 'Id_Xor-Cism'.

149. Ibid.

150. When asked about the reference, Mez replied that she intended reference to '*both* Baudrill[drilldrill]ard's usage [+ latin origins] + the game itself' (email 2010).

151. Helen J. Burgess, 'Introduction: Spaceworks', www.hyperrhiz.net/ hyperrhiz03/6-introduction/20-introduction-spaceworks.

152. Mary Anne (Mez) Breeze, '[Twitte]Reality_Fiction: 27/3/09', version.org/textuals/show/12.

153. Mary Anne (Mez) Breeze, 'Dam(B)a({N}D{a})Ged Poetry', delicious.com/TAGallery/EXHIBITION_solo.wurk. On the TAGallery exhibit site, Mez offers further valuable insight into her compositional practice:

　　1. a trigger is in[itially.e]volved [could b ph(r)ase/code strings,

a sensory cue, +/or information spark(s) via aggre(p)gation trawling].

2. a syn[es]t[ling]ax [dr(p)a(rsed)wn from my data absorbing at that time] then w[b]ra[nches]ps around the trigger.

3. then come the mechanisms employed 2 diffuse the impulse 2 use a [self]conscious authorial voice 2 construct the work. they include: /music [l(em)o(tive)ud] /fracture-tasking [like multitasking but demanding a type of staggered attention between consciousness projecting + an intense bombarding of stimulus: eg game.system.flitting or focused.stimulus. expansion+conTr[ee]action via data absorption].
[if i let myself create in my primary authorial voice via focused/ conscious attention normally associated with the creative process, all i'd produce is source code or dam(b)a({n}d{a})ged poetry].
[regulati(c)on(trol) of whe(n)re i'm creating is also a plus: this can be internally moderated (i tend 2 have outrageously expressive idea_channelling whilst trying to block out annoying passengers on long-range train trips, tho this could have 2 do with music pumped @ ear-bleeding lvls;)].

4. then comes the compiling: pie[r]cing 2gether via a streaming process of generating>constructing>analysing>re [m({h}ash)ix]constructing>m[f]ixing>optimizing etc blah until it has some type of internal consistency.

5. my wurks r never really finished; they kinda hang together in a faux_fixed state, rdy.4.the.next.incarnation (dam(b)a({n} d{a})ged).

154. Talan Memmott, 'E_RUPTURE://Codework'.'Serration in Electronic Literature', *American Book Review*, no. 22.6 (2001), talanmemmott.com/pdf/e_rupture.pdf.

155. www.secrettechnology.com/madethis/enemy6.html.

156. The other two games in the trilogy are *Evidence of Everything Exploding This* (www.secrettechnology.com/explode/evidence. html) and *Game, Game, Game and Again Game* (www. secrettechnology.com/gamegame/gamegame.html). These and other works produced by Nelson appear on his website, www. secrettechnology.com/.

157. Links to these sites are www.secrettechnology.com/poem_cube/ poem_cube.html and www.secrettechnology.com/ausco/ poecubic2.html, respectively.

158. Jason Nelson, 'I made this. you play this. we are enemies', www.secrettechnology.com/madethis/enemy6.html.

159. Ibid.

160. Ibid.

161. Less characteristic videos are embedded, signified by a star, but not compulsory to watch.

162. Nelson, 'I made this. you play this. we are enemies'.

163. Given the absence of predatory avatars or elements, restarting a level is the only penalty.

164. Nelson, 'I made this. you play this. we are enemies'.

165. Ibid.

166. Ibid.

167. Ibid.

168. Ibid.

169. Nelson is at his most critical on level 6 – titled 'Barrenland', defacing the Recording Industry of America (RIAA) website – where he unrepentantly castigates 'lawsuit as culture' folks' with lines such as 'the noteless swim/in waters bleak/and artless' and 'they sleep/on beds/filled with/the lunch/money of/100,000/poor teens', while giving players the opportunity to jump onto an Electronic Frontier Foundation platform and move on to the next level.

170. Point of clarification: as previously indicated, *I made this. you play this. we are enemies* has a fixed structure. Content appearing on each level is the same each time the game is played – no 'on the fly' permutation occurs.

171. Nelson, 'I made this. you play this. we are enemies'.

172. Ibid.

173. Ibid.

174. Ibid.

175. Ibid.

176. Ibid.

177. Ibid.

178. Ibid.

179. Ibid.

180. Ibid.

181. Ibid.

182. Ibid.

183. www.slippingglimpse.org.

184. Strickland's online publications include *Ballad of Sand and Harry Soot* (1999), *To Be Here As Stone Is* (1999) and *Errand Upon*

Which We Came (2001, with M. D. Coverly); stephaniestrickland. com/ contains her complete publication history.

185. For more discussion on *V: WaveSon.nets/Losing L'una* and *Vniverse*, see the authors' essay 'Making the Vniverse' (www.cddc.vt.edu/journals/newriver/strickland/essay/index. html) and my review of the book in *Electronic Book Review* (www.electronicbookreview.com/thread/electropoetics/ superdense?mode=print).

186. Videopages open in full-screen mode, a setting which can compromise video quality (minor pixillation, distortion). Multiple viewing options are offered ('high-rez video', 'scroll text'), as well as options to regenerate the current selection and links to previous and subsequent videos, enabling a process the Introductions describes as 'a round robin of reading'.

187. Stephanie Strickland, Cynthia Lawson Jarmillo, and Paul Ryan, 'slippingglimpse', www.slippingglimpse.org/.

188. Ibid.

189. Ibid.

190. Ibid., June 26, 2008.

191. Ibid. *slippingglimpse* also considers 'older capture technologies, such as harvesting plants for food and flax for paper'.

192. Ibid.

193. A mouse-controlled slider is also provided, so viewers are able to scroll in both directions through the text, and to pause it on the screen.

194. Strickland, Jarmillo, and Ryan, 'slippingglimpse'. June 26, 2008.

195. Ibid.

Notes to Chapter 5

1. Many works published as executable files do not require the WWW and capably function in offline circumstances.

2. *The Google Poem Generator* is currently, perhaps permanently, offline.

3. www.googlism.com/.

4. Formerly housed at www.leevilehto.net/google/google.asp.

5. I set my observations here in the past tense because *Googlism* does not access up-to-date information from Google. The version presently available essentially documents what the program did when 'live'; the information it draws from is long out of date: the same poems generated now would have appeared identically in 2004.

6. In the original version, users were given a choice via radio button to select 'who', 'what', 'where' or 'when'. *Googlism* today still seems to incorporate the feature, but it does not. Each button brings the same returns because of the status of the program, explained below.

7. Paul Cherry and Chris Morton, 'Googlism', www.googlism.com/.

8. *Googlism*'s database is large (1.7 million unique Googlisms in total), but frozen in time. According to the *Googlism* website, in early 2004 Google 'adopted measures to prevent Googlism. com from querying the Google servers to find new Googlisms'. Incidentally, if the input text is not found in the database, one can compile another type of list poem with returned results, such as: 'Sorry, Google doesn't know enough about reluctance yet./Sorry, Google doesn't know enough about obama yet.'

9. The project website stated, 'The techniques involved include Microsoft IIS5 web server and Active Server Page 3.0 scripting engine with VBScript.'

10. Because Google discourages its use for alternative purposes, and because code and coding methods change, authors appropriating the resource are required to reversion their works according to specific, sometimes changing, protocols. Commenting on the transitory aspects of tapping into resources such as Google images, Andrews, in an August 2010 email explains:

    > *dbCinema* and *Google Poem Generator* are examples of pieces that use web services. A web service is such that one requests info from it and the web service responds with the requested information. *dbCinema* requests info from Google concerning images associated with some language. Web services, like *dbCinema*, are in development. So how can they develop and yet still present a stable service that continues to work for programmers even as the web service changes stuff? The key notion here is that of the api, or application programming interface. An api defines a communications protocol. If you want information from the service, you phrase the request according to the rules of the api. Often there are ways to request information from services that do not adhere to the api protocol. The request may work today, but perhaps not tomorrow. Whereas the whole idea of setting up an api is so that if you request info according to the rules of the api protocol, your request will be successful today and

tomorrow. Even as the web service changes and grows. I'm currently changing *dbCinema* so that the requests it makes to Google for image information are according to Google's api. this should result in a *dbCinema* that works for quite some time.

11. Leevi Lehto, 'Google Poem Generator', www.leevilehto. net/google/about.asp. Interaction with the program was straightforward: users selected a pattern to work with, entered a 'Search string/Title', plus any 'Additional settings', and clicked a 'Get Poem' button. Additional settings included the language of pages to search, line length, number of lines, and, in the Sonnets section, the 'Rhyme scheme' (Italian or Spenserian). A link to background information, detailing rationale for these schemes, was provided for some patterns.

12. Most impressively, and perhaps most importantly, for sake of variation, the output presented for each search is unique. The language incorporated into the poems through the ASP programming varies from poem to poem and from pattern to pattern. A significant amount of verbal clutter usually emerges, and some examples are more readable than others and contain more interesting qualities. Having a diverse range of samples to choose from is beneficial to the process of appreciation of these, or any, generated poems.

13. See web.njit.edu/~funkhous/13states/. For a performance at the Thailand Media Art Festival (2006), I also produced three Flash works, combining image, sound and text: 'Feedback', 'How Fast Can a Zebra Run?', and 'Facts About Durians', whose texts are based in (edited) output from the *Google Poem Generator*.

14. Quite a few poets included Google as part of their composition process by this time. Digitally based writings of the Flarf poets embraced and exaggerated a practice that typically used the results of search strings to make poetry. Members of this collective originally set out to write awful poems, but by 2003 began to focus on using Google searches as a resource for the composition of poetry. In 2003, K. Silem Mohammad's *Deer Head Nation*, the first book of poems comprised entirely of Flarf, was published. Since then, the group, organized via a Googlegroup, has expanded and diversified, publishing many works and making numerous public presentations emphasizing humour and irreverence. Flarf has gained visibility (and notoriety) in the poetry community due to their unconventional approach. Sharon Mesmer, a member of the group, in a Feb. 2009 email

described her process, writing 'I put a phrase ... through Google, then I copy and paste the first five pages of search results in a doc, then go through the doc and make a poem out of that. I throw in some material from the actual emails, maybe some journal bits, and try to make it cohere a tiny bit (though not too much). I try to have it done in under fifteen minutes.' Speedy acquisition of raw, yet organized, text on the WWW is integral to Flarf poems, but is not the only aspect of the work. Such processes result, for Mesmer, in poems with titles such as 'Non-Pimpin' Huggy Bear', a bawdy poem that appeared on the Flarflist whose first stanza contains this passage:

> Time, being but a whisper on the lips of Mystery,
>
> affords us a slim chance of crossing a huggy bear
>
> who is singing kitty limericks about kitty implements
>
> with a pimp who is two Red Bulls past the event horizon
>
> as a tsunami of Quik-E Mart fantasy scenarios
>
> spins an ocean of pale flu pee that looks like Jarvis Cocker
>
> non-pimpin' the Unabomber's father
>
> selling sausage to my grandmother.

A coherent narrative, Mesmer peppers the poem with obscenity, literary and mainstream references, resonances of threads of other works (hers combined with search results), cobbled together with sensibility, with a point of address. Mesmer's poems, a series of partially non-sequitur, lively lines, curiously shape her humorous imagination. Use of technology in Flarf is very basic; the effort of the poem largely involves the writer's editing skills. Flarf uses technological processes to generate foundation texts, shaping them into a type of cyborgian expression – producing works as static text, suited for publication in print. A reader does not play a role in the production of the poem, does not supply any of the content; an author's efforts can be evaluated in the same manner as any other open form poem. An important example of how computer programming and communications networks have begun to influence directly page-based writers, Flarf shows how appropriation – one text building from the bones or meat of another – plays a role in one area of contemporary poetry. Readers interested in seeing more of Mesmer's work should seek out her chapbook *Annoying Diabetic Bitch* (Combo Books, 2008) or explore her blog, *? dubious labia !* (dubiouslabia.wordpress.com/). Other important Flarf publications include: *Deer Head Nation* (Oakland, CA: Tougher Disguises, 2003); a Flarf feature in

Jacket 30 (jacketmagazine.com/30/, 2006); a Flarf anthology is forthcoming (Edge 2012). In 2007, several years after the initial appearance and publications of Flarf, a second group promoting the use of Google in the composition of poetry, Poegles, was launched. The group's website (www.poegles.com) contains an explanatory section, 'How to Make a Poegle,' outlining a four-step compositional process that conceptually resonates with my description of using the *Google Poem Generator* (above): '1. Enter a phrase into a search engine; 2. Copy the interesting results; 3. Paste onto a single page; 4. Edit to make a poegle.' While Flarf poets impose no rules or requisite processes, the poegles group utilizes a distinct form and process that technically equates to what Flarf does. Aesthetic differences are glaring, however: a poegle intends to refine the search results poetically into a sensible utterance, whereas Flarf remains discursive and preserves content intended to annoy or irritate the reader. Further, *Poegles* sponsors a monthly contest, offering $100 and publication for the best poegle poems submitted; the site also features a 'Collection' section that contains selections from the book *Poegles: A Short History and Collection* (Justin Hendrix and Dave Gunton, eds).

15. *Google Poem Generator*, for me, never issued a completed poem, so I developed a process for preparing focused works; I used the program to generate a series of texts using different patterns (usually combination of sonnets, couplets and collage), and then sequentially pasted together the output in a Microsoft Word file. I closely edited the content of these files so as to omit any ordinary or cliché travel agent (i.e. tourist-oriented) rhetoric, and eliminate information that did not contribute to the content and sound of the poem taking shape. The elaborate editorial process was necessary due to the enormous volume of non-poetic and commercially-oriented text. After an editorial pass or two, I printed and re-edited files. Whenever I made changes to an electronic file, letters and words removed were replaced by the same amount of blank spaces, rendering a visual component on the page.

16. C. Funkhouser, *Dragon Fruit,* 2006. Poem from '13 States of Malaysia'.

17. The first stanza retains only a single word from the original, i.e. 'in', completely unrelated to the topic. I edited out as much as eighty per cent of the original output (which was usually 10-15 pages) in most examples. Since output from the *Google Poem Generator* is (with the exception of the 'collage' pattern) aligned justified to the left margin, someone who reads the '13 States

of Malaysia' poems can gauge how much text is removed by recalling that the initial output was tightly packed together. As a student of Charles Olson's 'Projective Verse' (and its assertion of the poem as 'a field') with great respect for poetic senses of language and sound, musicality brought to the task involves ear as well as eye and intuitive mind. I spent months creating and revising the 13 poems for the Fulbright project, leading some colleagues to ask 'wouldn't it take less time and be easier if you just wrote the poems yourself?' In fact, by fusing small poems I did write – combining whatever special views I had of the physical and cultural landscape – with what I was able to observe, learn and craft from the network, harnessed by a poetic programmer. This brought an objective focus of content, form and shapes to the poems I would not have accomplished on my own. I liked what I could do with these procedures: net-ordered topical leaps, infusion of Bahasa language, unexplainable contrasts and discursive narrative issued surreal peripatetic wanderings. Fusing subject and WWW, the program achieved results my own sensibilities would not have intuitively derived. *Google Poem Generator* does not write a finished poem (although it may, especially for those interested in Codework) but rather offers the framework for a possible poem, which the user can further modify. For a purist, the various tags and de-contextualized, non-grammatical materials are part of a Google poem and should be appreciated and even evaluated as such. However, users of this or any generator are not obliged to preserve all parts of the output.

18. Funkhouser, *Dragon Fruit*.

19. This theme is cryptically encapsulated in the poem's opening acrostic:

    Downright

    Ridged

    Agriculture

    Growing

    Over

    Nation

    Farout

    Rouge (rogue)

    Utility

    Integrates

    Texture.

20. 'Any sufficiently advanced information is indistinguishable from noise' – Christian Bök. Twitter, April 12, 2009.

21. Lehto describes *Google Poem Generator* as 'a collective artistic celebration of, and reflection on, the great revolution in the structures of information delivery and retrieval that Google is so much part of.' Using what the program delivered as a basis for a poem resulted in one of my most versatile works yet – presented on pages (of magazines), stages (three-channel video/animation installation with sound and performance), and in galleries (large posters). My background as a poet, trained to compose with crafted, resonant language, and my software – tools provided to help with the task – made this work possible.

22. www.vispo.com/dbcinema/.

23. Jim Andrews, 'dbCinema', vispo.com/dbcinema.

24. The online version enables use of online resources/directories; Google, Yahoo, or a combination of the two (GooYa, YaGoo) are prescribed by Andrews, but the viewer (in the 'image settings' section of the 'Gear Shift') can specify any online directory. In the offline version, in addition to the net image search the viewer has the ability to select images from her/his own computer, and can do a 'domain search' to acquire images from alternative online sources.

25. Andrews developed the advanced version of the program for commercial purposes and it was not publicly available until 2010.

26. Jim Andrews, 'Kandinsky 3: dbCinema Images Using the Flash Brush,' vispo.com/dbcinema/kandinsky3/intro/.

27. Jim Andrews, Vispo, vispo.com/dbcinema/huth/index.htm.

28. Images *dbCinema* obtained to create this were drawn from a Google image search of 'Geof Huth'.

29. Note: the user also has the option to set an 'Underlying BG Stage Colour'.

30. Andrews, 'dbCinema'. In the most recent, 'Interactive online dbCinema' version, a new brush, 'Colour', has been added.

31. Jim Andrews, email, Feb. 2009.

32. Andrews, 'Kandinsky 3: dbCinema Images Using the Flash Brush'.

33. www.serandour.com/tue-moi.htm.

34. For more on this approach, see discussion of Frédéric Develay's 'En toutes lettres' in *Prehistoric Digital Poetry*, 132–3.

35. Eric Sérandour, 'Tue-Moi', www.serandour.com/tue-moi.htm.

36. Ibid.

37. Ibid.

38. Talan Memmott in a personal conversation in Fall 2008 explained that certain parts of his landmark work *Lexia to Perplexia* (2000) have begun to cease to perform as designed with recent browser versions. In order for functionality to be restored, Memmott would have to return to the code and make updates, a chore he is unwilling to do.

39. Philippe Bootz, who published 'Tue-moi' in *Alire* 11, confirms that the address never existed in his essay *'Alire*: Une Expérimentation de *Poésie* Informatique en Lecture *Privée'* (www.isea-webarchive.org/mmbase/attachments/36548/16_bootz.pdf).

40. Note: According to Sérandour's dossier, 'Tue-moi' is a response to Christopher Tarkos' *Caisses* (1998), which is available on the WWW via www.sitec.fr/users/akenatondocks/DOCKS-datas_f/collect_f/auteurs_f/S_f/serandour_f/anim_f/secaisses_f/caisses.htm. April 3, 2009.

41. motorhueso.net/dcr/synonymovie/synonymovie.htm.

42. V. Flusser, A. Ströhl and E. Eisel, *Writings* (University of Minnesota Press, 2002). Each experiment featured on Tisselli's website (motorhueso.net) contains visual attributes.

43. Eugenio Tisselli, email, Feb. 2009. A movie's development through time 'is relational,' writes Tisselli, 'the movie's current frame is directly related to the previous one in a semantic way,' initiated from the user's theme.

44. An open-source scripting language, PHP creates dynamic web pages by embedding html commands that generate html on the server, which is sent to the client. According to *Wikipedia*, PHP has 'evolved to include a command line interface capability and can be used in standalone graphical applications.' (PHP). *Synonymovie* is written in Lingo, the scripting language employed by Adobe Director (software), and functions on the WWW by use of the Shockwave plugin device, as do Flash animations.

45. thesaurus.reference.com. Note: in 2011 Tisselli changed the image source location to Flickr.

46. In a Feb. 2009 email, Tisselli notes that *Synonymovie* requires extensive maintenance because he does not use the Google API to get the images, but instead uses a technique called 'screen scraping', which is based in parsing the html coming from the Google image search results and mining the image data. Tisselli writes, 'You can imagine that this 'inner' Google code is subject to periodic changes (different servers, different ways of coding …) so every time Google makes a change in its html code, I have to adapt *Synonymovie* accordingly. Needless to say, the only way to know that a change must be made is by realizing that *Synonymovie* is not working properly, so I check my stuff almost every day… but fortunately, I only have to do changes about once or twice a year….'

47. Entering the phrase 'flarf' into *Synonymovie* roughly twenty times over two days, five different single screen movies were generated multiple times; one had no visual content.

48. Eugenio Tisselli, 'Synonymovie', www.motorhueso.net/dcr/ synonymovie/synonymovie.htm.

49. Close readers will suspect the jump from 'make hazy' to 'assemble' results from the antonym function.

50. Given the limits of the entries of the thesaurus, some repetition occurs, occasionally taking form as a type of feedback loop in which paired content shuffles and alternates on the screen.

51. Tisselli, 'Synonymovie.'

52. Viewers who want to revisit the output will need to activate a video capture device. I use digital video to preserve the materials as research, although a software program such as Camtasia or Jing could also be used effectively. Best practice for engaging with *Synonymovie*, in any event, is to take time, at different hours and on different days, to see how different words perform, how they become verbally and visually activated when entered into the program and brought forth from the WWW. The viewer has the liberty to begin anywhere, and enter any concept for consideration. I have watched very profound animations based on the words 'poetry' and 'homeless', both of which intoned commentary on creative and cultural structures in unpredictable ways.

53. turbulence.org/Works/without_a_trace/.

54. Zellen's ongoing work, *Urban Fragments* (2009–10) is an up-to-date representation of her exploration in this area (see urbanfragments.net). *Seen Death* (2007, www.jodyzellen.com/ oog4/index.html) also reflects similar attributes, i.e. combining newspapers, animated text/image and pop-ups.

55. Jody Zellen, 'Without a Trace', turbulence.org/Works/without_a_ trace/. She describes her 'quirky and purposely imprecise' hand-drawing style as 'a distilled interpretation of the news'.

56. Ibid.

57. Jody Zellen, *Digital Poetry Performance,* 2009. E-Poetry 2009:

An International Digital Poetry Festival. In addition to using the automated technique in *Without a Trace*, Zellen uses it as a component in her 2010 works *The Lines of War* (www. urbanfragments.net/war_grids/) and *Lines of Life* (www. urbanfragments.net/drawnwar/), as well as in 2007's *Empty thoughts for real life* (www.jodyzellen.com/comic/index.html) and *Seen Read + Drawn* (www.jodyzellen.com/drawings_ani.html) and other titles.

58. Zellen also uses the *New York Times* as a point of departure in her 2006 work *All the News thats Fit to Print* (www. allthenewsthatsfittoprint.net/).

59. In response to a question I posed to her about the activity involved with such an integrated composition, Zellen writes in an Aug. 2009 email she, 'created an archive that contained 365 images of each type (traced drawings and altered comic). The code was written in such a way that these are only used once and they are randomly called upon. The RSS feed is always changing as is the news image that the computer "traces"'.

60. Zellen, *Digital Poetry Performance*.

61. The same effect, minus the newsfeeds being up-to-date, can be seen by clicking the 'Random Version' button on the project's archive page.

62. Zellen, 'Without a Trace'.

63. Under the random version link viewers can refresh the page repeatedly to collage and re-collage a different set of archived elements. In the active version of *Without a Trace*, only the headline was live; in the presently archived version, the headline text ('Economy Grew at Rate of 2.4% in Quarter') is the only element that does not (will not) change when the page is refreshed.

64. Zellen, 'Without a Trace'. See turbulence.org/Works/without_a_trace/screenshots.php?month=08 for the archives for this period.

65. To view Tisselli's *Dadanewsfeed*, see motorhueso.net/newsfeed/index.htm.

66. Zellen, 'Without a Trace'.

67. Jody Zellen, email, August 2009. Zellen explains her process in this section of *Without a Trace* as follows: 'I copied out the words from the original comic and saved them in an archive. There are about 150 words that are randomly chosen to fill the three lines of words.'

68. Zellen, 'Without a Trace'. The following lines, which perhaps

better reflect the way a fixed set of lines are slotted into Zellen's construction, are taken from the slides archived on the *Without a Trace* website: 'blow/will/found'; 'forget it./got/knowing'; 'look at/the/early'; 'only/look at/sign'; 'whatever/my/look at'; 'sadly./ choose/left,'; and 'being/something/left,'.

69. The archived screenshot, in addition to the Jihadist headline, includes the texts: 'Whatever/My/Look At' (from comic); 'A big industry, a bigger responsibility' (Zellen); no text appears in the image (a cityscape). In both examples, two empty comic balloons are present.

70. Jody Zellen, 'All the News Thats Fit to Print', www. allthenewsthatsfittoprint.net/.

71. Zellen, 'Without a Trace'. Every morning she manually made an image of the site and uploaded it.

72. Philippe Bootz, 'Désir Insuffisant', www.sitec.fr/users/ akenatondocks/DOCKS-datas_f/collect_f/auteurs_f/B_f/BOOTZ_F/ TEXTES_f/Desir_Insuffi.htm.

73. John Cayley and Giles Perring, 'Imposition', programmatology. shadoof.net/index.php?p=works/impose/impepoetry.html.

74. Philippe Bootz, email, Jan. 2009.

75. John Cayley, 'Program Notes for Imposition', programmatology. shadoof.net/works/impose/impositionProgNotes.pdf.

76. Ibid.

77. Ibid. In comments on *translation*, Cayley reveals the work's main source text as an early essay by Walter Benjamin ('On Language as Such and on the Language of Man').

78. The term is used to describe 'textual morphing based on letter replacements' (see *Prehistoric Digital Poetry*, 229).

79. Cayley, 'Program Notes for Imposition'.

80. As stated in a diagram included with Cayley's description of the work, a 'servlet listens for states and rebroadcasts them' (Notes).

81. N. Katherine Hayles, *Electronic Literature: New Horizons for the Literary* (Notre Dame, Ind.: University of Notre Dame Ward-Phillips, 2008, 151).

82. Theodor H. Nelson, *Geeks Bearing Gifts: How the Computer World Got This Way* (Sausalito, CA: Mindful Press : Distributed by Lulu.com, 2008, 179).

Notes to Chapter 6

1. Marjorie Perloff, *Unoriginal Genius: Poetry by Other Means in the New Century* (Chicago: The University of Chicago Press, 2010), xi.

2. The matter of dissemination is also interesting because the WWW unquestionably alters methods of circulating literary materials, making it more of a creative process.

3. Theodor H. Nelson, *Geeks Bearing Gifts: How the Computer World Got This Way* (Sausalito, CA: Mindful Press : Distributed by Lulu.com, 2008), 71. For Nelson, directory structures used on the WWW, the particulars of html coding and the use of CSS present sometimes unforgiving chains of command and impenetrable barriers leading to the isolation of, and disruption within, texts and publications that could be more versatile and available. A simpler, more effective approach to network functionality would demystify the apparatus, and thus encourage participation on a compositional level.

4. Ibid., 75.

5. Ibid., 122–3. To underscore the Xerox scheme's dominance, Nelson switches the name for graphical user interface (GUI) to PUI (PARC user interface).

6. Ibid., 167.

7. Ibid., 130.

8. J. Rasula, 'From Corset to Podcast: The Question of Poetry Now', *American Literary History* 21, no. 3 (2009).

9. In addition to *Googlism*, *Google Poem Generator* and *Erika*, some of my own works and works by my students – made with JavaScript and Java applets – also no longer function.

10. Deena Larsen, email, 23 April 2009.

11. C. Anderson and M. Wolff, 'The Web Is Dead. Long Live the Internet', *Wired Magazine* (2010).

12. 'Internet Archive: Wayback Machine', www.archive.org/web/web.php. The Wayback Machine is 'a digital time capsule created by the Internet Archive non-profit organization.... This service allows users to see archived versions of web pages across time – what the Archive calls a 'three dimensional index'. (see 'Wayback Machine', en.wikipedia.org/wiki/Wayback_machine.).

13. C. Funkhouser, 'Encapsulating E-Poetry 2009: Some Views on Contemporary Digital Poetry', dichtung-digital, www.brown.edu/Research/dichtung-digital/2009/Funkhouser/index.htm.

14. Philippe Bootz, 'Archipoenum', code.google.com/p/archipoenum/. Bootz' presentation included a demonstration of the current state of the tool, an open-source work-in-process delivered as a Firefox plug-in.

15. The Electronic Literature Organization is involved with the preservation of information works of digital literature through their *Electronic Literature Directory* (directory.eliterature.org/), which features an edited and annotated bibliography of electronic literature. Another initiative, the ELMCIP Knowledge Base, produced by a collaborative research project titled 'Electronic Literature as a Model of Creativity and Innovation in Practice', is presently building a cross-referenced, contextualized information database about authors, creative works, critical writing and practices (elmcip.net/knowledgebase).

16. For example, in 2007 an extraordinary team of artists – Jim Andrews, Geof Huth, Lionel Kearns, Marko Niemi and Dan Waber – worked together to re-engineer bpNichol's 1980s Apple BASIC program *First Screening* (www.vispo.com/bp/introduction.htm).

17. Nelson, *Geeks Bearing Gifts: How the Computer World Got This Way*, 187.

18. Ibid., 25.

19. N. Katherine Hayles, *Electronic Literature: New Horizons for the Literary* (Notre Dame, Ind.: University of Notre Dame Ward-Phillips, 2008), 13.

20. Nelson, *Geeks Bearing Gifts: How the Computer World Got This Way*, 92. By this he means movies that enable multiple narrative directions guided by a viewer's choices.

21. Ibid., 57.

22. Talan Memmott, 'Beyond Taxonomy', in *New Media Poetics: Contexts, Technotexts, and Theories*, ed. Adalaide Kirby Morris and Thomas Swiss (Cambridge, MA: The MIT P, 2006), 302.

23. Ibid., 301.

24. In the previous volume I worked with Spanish, French, German and Portuguese translators in order to provide a more broadly focused historical study. In significant contrast, here I chose to discuss only works prepared in English, largely by North American artists. Because digital poetry is practised globally, a comprehensive study of contemporary works would indeed include works in many languages. At this point, I defer to scholars competent in the respective (and respected) languages to penetrate the intricacies of these works.

25. Carolyn Guertin, 'Handholding, Remixing, and the Instant Replay: New Narratives in a Postnarrative World', in *A Companion to Digital Literary Studies*, ed. Susan Schreibman and Ray Siemens (Malden, MA: Blackwell, 2007), 239.

26. aND has also produced a range of static works, made interactive works available for download, and with collaborator Maria Damon engineered numerous hypertext titles (such as *pleasureTEXTpossession*, *Literature Nation with Hyperpoesy*, *Semetrix* and *eros/ion*). A complete index of aND's online work is available via his *JOGLARS\crossmedia beliefware* site.

27. Cayley has recently created (2011) another clock, using Processing, without photographic properties; see, programmatology.shadoof.net/applets/epigraphicClock/.

28. Modes of composition used by works highlighted in chapter 4, in addition to Adobe's multimedia programs Flash and Director, include MySQL (database), JavaScript, Processing and Java (applet).

29. Hayles, *Electronic Literature: New Horizons for the Literary*, 8.

30. Ibid., 23.

31. 'The Electronic Literature Collection, Vol. 2', collection.eliterature. org/2/. Net-poetry titles include Tisselli's *Dadanewsfeed* and Ton Ferret's *The Fugue Book*. J. R. Carpenter's *In Absentia* and Christoph Benda's *Senghor on the Rocks* incorporate Google maps into their writings; Caitlin Fisher's *Andromeda* augmented reality piece (introduced below) and documented in the publication, relies on a webcam to function.

32. C. Funkhouser, 'E-Poetry 2003: An International Festival of Digital Poetry', (2003), www.wepress.org/E_Poetry_2003.pdf.

33. Caitlin Fisher, 'Andromeda. Augmented Reality Poem', Authoring Software, www.well.com/user/jmalloy/elit/elit_software. html#caitlin.

34. Funkhouser, 'Encapsulating E-Poetry 2009: Some Views on Contemporary Digital Poetry'.

35. This software was developed by researchers working in Fisher's lab at York University.

36. Christine Wilks, 'Augmented E-Poetry at ELO_AI', *Netpoetic* (2010), netpoetic.com/2010/06/augmented-e-poetry-at-elo_ai/.

37. Mary Anne (Mez) Breeze, '_Feralc_', netwurker.net/.

38. Mary Anne (Mez) Breeze, 'Video: Bruce Sterling's Keynote – at the Dawn of the Augmented Reality Industry', site.layar.com/company/blog/

video-bruce-sterlings-keynote-at-the-dawn-of-the-augmented-reality-industry/. In an email, mez writes in _feralC_, 'I'm using both QR codes [see one of the characters here: twitter.com/qreada] + also a synthetic pop-up AR book for Episode 2.'

39. Oswald de Andrade, 'Anthropophagist Manifesto', *Latin American Literature and the Arts* no. 51, 65. See www.agencetopo.qc.ca/carnages/manifeste.html for a translation by Adriano Pedrosa and Veronica Cordeiro. For more background on anthropophagy see João Cezar de Castro Rocha, 'Brazil as Exposition', *Ciberletras* 8, no. 8 (2002), www.lehman.cuny.edu/ciberletras/v08/rocha.html.

40. Elizabeth Bishop, ed. *An Anthology of Twentieth-Century Brazilian Poetry* (Hanover, NH: Wesleyan U P, 1972), 11.

41. Ibid., 13.

42. Raul Bopp and Poty, *Cobra Norato E Outros Poemas* (Rio de Janeiro: Civilização Brasileira, 1973), 16.

43. Marcus Salgado, email, April 2007.

44. A discussion about the manifestations of anthropophagic poems in Brazil would address works by de Campos, Décio Pignatari and other concrete poets, as well as other historical figures (e.g. Flavio de Carvalho) and younger artists who practise with intent today. Here I would like to thank both Lucio Agra and Marcus Salgado for bringing relevant works by historical and contemporary artists to my attention, and for our ongoing dialogues, which have substantially contributed to my discussion of this topic.

45. References to concrete poetry are far from uncommon in dialogues regarding the influence of literature on new media productions: concrete poetry has been cited as an influence on computer poems since the 1970s. See R. W. Bailey, *Computer Poems* (Potagannissing Press, 1973) and Carole Spearin McCauley, *Computers and Creativity* (New York: Praeger, 1974). Bailey writes that in graphical computer poems 'concrete poetry is reflected with a computer mirror' (np); McCauley writes computerized graphical poetry 'resembles, or perhaps grew from … "concrete poetry"' (115). More recent books on the subject, such as Loss Pequeño Glazier's *Digital Poetics* (2002) and Brian Kim Stefans' *Fashionable Noise* (2003), as well as essays by Friedrich Block and Roberto Simanowski, discuss the relevance of concrete poetry to the development of digital poems.

46. Campos, Augusto de, Personal interview, 2005. In a Nov. 2006 email exchange regarding the ritualistic elements in concrete poetry, de Campos writes, 'We viewed Anthropophagy as

an anthropologic metaphor, nurtured in Freud, Nietzsche, Lévy-Bruhl and Bachoffen (from whom he took the theory of ancient Matriarchy, that would have preceded Patriarchal society, associated with authoritarian monarchies and private propriety)…. The brainstorming in which we were engaged, in a Poundian way ('paideuma', 'the age demanded'), tried 'to gather from the air a live tradition', reading in several languages as only barbarians do to arrive at the selective choice MALLARMÉ-JOYCE-POUND=CUMMINGS was surely linked to the Oswaldian cultural ANTHROPOPHAGY'.

47. Marjorie Perloff, *Radical Artifice: Writing Poetry in the Age of Media* (Chicago: University of Chicago Press, 1991), 3.

48. Ibid., 19.

49. Bill Seaman, 'Recombinant Poetics', in *Media Poetry: An International Anthology*, ed. Eduardo Kac (Chicago & London: Intellect Books, 2007). Seaman's recombinant poetics explores the generation of 'mutable poetic contexts' (173); his essay on the subject appears in *Media Poetry* (157–74), 157.

50. Charles Bernstein, 'De Campos Thou Art Translated (Knot)', epc. buffalo.edu/authors/bernstein/essays/de-campos.html.

51. D.J.S.T.S. Kid and P. Lunenfeld, *Rhythm Science* (MIT Press, 2004), 25.

52. While inclined to focus on poems featured in the case studies, my own work often cannibalizes a limited set of letters to produce anagrammatic poems around a given theme. The following passage describes exactly how cannibalistic contexts played a role at a 2009 performance in Bergen, Norway: both visual components co-opted cannibalistic network devices. For projected poems (also spoken at the outset), I compile then edit phrases made with the Internet Anagram Server to create lines of customized expression. From the 1,181 results lines I received from three query phrases ('Norway delicately', 'Bergen light' and 'Bergen lightly'), 41 lines are chosen to create a text file (Bergen light II) that becomes the basis for the animated poem. Through this I offer a slow, strange, personal, calculated speculation (now public statement) with words, their sounds, and interactions. The accompanying soundtrack is original, but has a relevant back story: I recorded text (words) and remixed them twice, into separate tracks; both tracks, at different speeds, stretched the words to a point beyond grammatical recognition. On a third audio track, in harmony with the distorted text, I dubbed a one-take composition on a one-string guitar that uses for its body a Coca-Cola can. For the final mix, I removed the first two audio tracks altogether and add rhythmic effects to the sound. Myriad approaches are used in these compositions. The avant-machinimatic companion in the projected montage is created with Jim Andrews' *dbCinema*. I created five new brushes (Bergen, delicate-poetry, espenaarseth, Bergen_history and lightly), each given both unique and common attributes. These brushes are arranged on a playlist, which also included a brush titled cannibalism previously devised by Andrews and me for E-Poetry 2009. Each brush configures verbal and visual information in a circular pattern. Some of the relations in the brushes are straightforward, others not. Bergen matches the name of the city with results of an image search on the city; Bergen_history matches the word 'history' with images of 'Bergen history'; cannibalism displays an animation of the word and images culled from a search of the word. To add poetic intent, however, delicate-poetry matches the word 'poetry' with the image results for 'delicate'; lightly matches 'lightly' with images tagged with the word 'dreaming'; and espenaaresth matches the word 'cybertext' with images of Espen.

Two of the poems spoken during the performance were made as a result of my participation in the Flarf collective. Beyond particular use of network technology and editing processes, Flarf is often, if not overwhelmingly, also based on discussions, exchanges and utterances that occur on the group's enormously active online discussion forum. These particular poems directly responded to posts I'd read by others. 'Scared' was a response to a Ben Friedlander gripe about the overuse of scare quotes, and 'Psychographic' was the last of a series of poems referencing Michael Jackson appearing on our forum during the weeks after his death. While the last few lines of the reading came from a Facebook comment box exchange between Tisselli, Adam Saponara and myself, the bulk of the text I read during the performance was drawn from a manuscript titled 'You are, Therefore I am', excerpts from a serial narrative I published on Twitter (see twitter.com/ctfunkhouser).

Each line of the work originates from a sentence made with Charles O. Hartman's program *PyProse* (oak.conncoll.edu/cohar/Programs.htm). For this project, I tested (with successful results) Hartman's idea (stated in *Virtual Muse*) that his generator 'could be treated as a first draft writer' (83). The program is open source – I have made a few adjustments to the code, and every morning I used it to produce some lines then fashioned via close editing – often to conform to the platform's constraint – into Twitter posts. In addition to other benefits, using the program brings words I never use but like, as well as strange modes of logic out of the blue that I – in turn, in a much disciplined way – force myself to respond to and shape into meaningful expression. This work is thus generated, and then filtered through my mind and sensibilities.

53. Bernstein, 'De Campos Thou Art Translated (Knot)'.

54. Ibid.

55. Ibid.

56. Millie Niss, 'The Electronic Muse', collection.eliterature.org/1/works/niss__oulipoems.html.

57. 'The Shannonizer', www.nightgarden.com/shannon.htm

58. Another interesting generator, *Gnoetry*, uses texts by specified authors to generate new poems. The *Gnoetry* project consists of *Gnoetry Daily*, a daily blog presenting 'statistically analyzed' texts (gnoetrydaily.wordpress.com/) and a website containing individual 'Beard of Bees Publications' (www.beardofbees.com/gnoetry.html). A number of compelling poems have been produced by completing a statistical analysis of pre-existing texts, although the program used to generate the poems itself is not yet publicly available. Coincidentally, a recent Beard of Bees project, *a light heart, its black thoughts*, by Gnoetry and Eric Scovel (#60: March, 2009) is a sonnet cycle that uses Conrad's *Heart of Darkness* as its input text (as does *Erika*). Other *Gnoetry Daily* posts use J. S. Mill's, *Utilitarianism*, Stevenson's *Treasure Island*, Agatha Christie's, *The Mysterious Affair at Styles*, H. G. Wells, *The Time Machine* and *Indian Erotica* as input texts.

59. Niss, 'The Electronic Muse'.

60. 'The Shannonizer'.

61. Ibid., 8 July 2008.

62. In 2006, while living in Malaysia, I was approached by a professor who had heard my lectures on digital poetry. His attention to my research arose from our common interest in generated poetry. He proposed a surprising collaboration: that we use AI (Artificial Intelligence)-related Neural Networks information processing techniques to write new Rumi poems. What we needed to do in order to make this happen was to build a vocabulary, then teach the machine how to speak like Rumi. It was a refreshing, alternative approach to the task, and one that might work if we were crafty enough. But due to physical separation (I returned to the United States), unfortunately, nothing ever happened with the scheme we concocted. Eventually dead poets will continue to

feast on their own media, contrived in the afterlife; poets will find ways to continue creating works despite physical absence from the planet.

63. The July/August 2009 issue of the esteemed monthly journal *Poetry* contained a feature on Flarf.

64. A program included on *SYNTEXT* recreates one of the earliest generators, Nanni Balestrini's 'TAPE MARK' (1961). Balestrini's work appropriates texts by Lao Tzu (*Tao Te Ching*), Paul Goldwin (*The Mystery of the Elevator*) and Michihito Hachiya (*Hiroshima Diary*). The program combines and constructs chains of words from these passages, but as a result of the inclusion of Hachiya's text ultimately and unavoidably portrays a scenario of nuclear disaster, albeit always through new perspectives rendered by the other two texts. Pedro Barbosa and Abilio Cavalheiro's *SYNTEXT*, an anthology of programs that appeared in the hypermedia journal *Alire* (1994), provides numerous examples of this dynamic and perspective on textuality. Poems and prose poems made by the programs include various rules and process – through a type of virtual consumption – words from databases to create new texts. The first experiments in computer poetry in the United States, Emmett Williams' 'Music' and 'The IBM Poem', clearly embodied anthropophagic aesthetics as had Lutz's stochastic texts. In 'Music' (1965), Williams used an IBM 1070 to identify the 101 most common words from Dante's *Divine Comedy*, and used them to create a series of computer poems. Williams borrows a condensed verbal framework from Dante, which is mechanically represented into lines that diminish, in relation to the number of times they appear in *Divine Comedy*, until a single word remains. In 'The IBM Poem' (1966) twenty-six words are randomly chosen from a dictionary and each is associated in a list with a letter of the alphabet to form lines; the letters of words in one line are then permuted to make subsequent lines. Several other American artists, such as Jackson Mac Low, John Cage and Charles Hartman, subsequently cultivated similar electronic approaches to composition. A fascinating program that 'read' and transformed (recomposed) a text provided by the user, who also sets various parameters, TRAVESTY, was developed by Hugh Kenner and Joseph O'Rourke in the early 1980s. Both Cage and Mac Low very ritualistically used appropriative, chance and computational methods in their work before having access to computers. Cage experimented extensively with aleatoric process (see *Prehistoric Digital Poetry* 64-5), as did Mac Low (67–8).

65. Stefans, Brian Kim, 'The Dreamlife of Letters'. collection.

eliterature.org/1/works/stefans__the_dreamlife_of_letters/
dreamlife_index.html.

66.   Strehovec, Janez, 'Text as Loop: On Visual and Kinetic Textuality'.
(2003), www.thefreelibrary.com/Text+as+loop%3A+on+visual+an
d+kinetic+textuality.-
a0113683509.

67.   When Nelson made a presentation at an NJIT Web design class
in 2008, he instructed students to appropriate .fla (i.e. Flash
production) code from resources such as Flashkit (www.flashkit.
com), explaining that he had used code from this and other
resources to prepare some of his works.

68.   Roberto Simanowski, 'Understanding Text That Moves: Two Close
Readings', *E-Poetry 2009: An International Festival of Digital
Poetry* (2009).

69.   Ibid. Examples in Simanowski's lecture included *The Messenger*
(Paul de Marinis, 1998/2005), *The Complete Works of W.S.* (Caleb
Larsen, 2008) and *bit.falls* (Julius Popp, 2006).

70.   Philippe Bootz, 'Discussion Comment E-Poetry 2009: An
International Digital Poetry Festival'.

71.   Funkhouser, 'Encapsulating E-Poetry 2009: Some Views on
Contemporary Digital Poetry'.

72.   Offline cannibalistic sampling of all sorts pervaded the Festival.
In *How to hear a sentence*, Marisa Plumb and Jonathan
Ben-Meshulam made screen texts by collaboratively using
extractions – most significant or ambiguous words – from a
written text about language and communication, writing them
into each other's texts, with the result of passages such as
'I am advocating a Lean Hypothesis about reality and a Lean
Alternative to our materialistic culture.' In the pop-art digital
poem 'Popup', made entirely with pop-up windows, Gerard
Altaió appropriates samples of pop songs which thematically
contextualize his minimalist work. In making 'Tokyo Garage',
Scott Rettberg cannibalizes the code from Nick Monfort's
'Taroko Gorge' to make his own generated poems. The sound
and image content of Alfred Marseille and Jan Baeke's piece
'What we had has not yet been' entirely cannibalizes found
film footage. A partially linear, partially non-linear, fragmentary
commentary on (or perhaps question about) domestic culture,
the poem begs the viewer to read into the combinations
presented by work to determine the message perhaps indicated
by the title.

73.   'Netpoetic', netpoetic.com/.

74. 'Netartery', netartery.vispo.com/.

75. Memmott, 'Beyond Taxonomy', 296.

76. C. Funkhouser, *Prehistoric Digital Poetry: An Archaeology of Forms, 1959–1995* (University of Alabama Press, 2007), 240.

77. Janez Strehovec, 'The Poetics of Elevator Pitch', *E-Poetry 2009: An International Digital Poetry Festival* (Barcelona, Spain).

78. Ibid., 6.

79. 'Attention Span', en.wikipedia.org/wiki/Attention_span.

80. Hayles, *Electronic Literature: New Horizons for the Literary*, 117.

81. Ibid.

82. David Glenn, 'Divided Attention', *The Chronicle of Higher Education* (Feb. 28, 2010), chronicle.com/article/Scholars-Turn-Their-Attention/63746/.

83. My interests compel me to make videotapes to capture streaming or never-seen-the-same-way-twice works. I would not, however, expect the average reader to do this.

84. Memmott, 'Beyond Taxonomy', 303.

85. Allen Ginsberg, *Cosmopolitan Greetings: Poems, 1986–1992* (New York: HarperCollins, 1994), 12.

86. Nathan Brown, 'The Function of Digital Poetry at the Present Time', newhorizons.eliterature.org/essay.php@id=11.html.

87. Robert Coover, 'Afternoon Seminar' (Albany: NY, 1996).

Notes to Epilogue

1. See www.youtube.com/ctfunkhouser#p/a/u/1/LDQZ6epXgcQ for documentation of the song's performance.

2. MIDI is an acronym for Musical Instrument Digital Interface, whose protocols enable electronic instruments and computers to communicate and synchronize with each other.

3. See epc.buffalo.edu/e-poetry/2011/ for event details.

Bibliography

Aarseth, Espen, *Cybertext: Perspectives on Ergodic Literature*. Baltimore: Johns Hopkins U P, 1997.

Amerika, Mark, 'Grammatron'. www.grammatron.com.

aND, mIEKAL, 'Seedsigns for Philadelpho'. english.umn.edu/joglars/SEEDSIGN/index.html.

—'Mesostics for Dick Higgins'. english.umn.edu/joglars/mesostics/index.html.

—'Spidertangle'. www.spidertangle.net/home.html.

Anderson, C., and M. Wolff, 'The Web Is Dead. Long Live the Internet'. *Wired Magazine* (2010): 118–27.

Andrade, Carlos Drummond de, Rafael Alberti and Mark Strand, *Looking for Poetry: Poems by Carlos Drummond De Andrade and Rafael Alberti and Songs from the Quechua*. New York: Alfred A. Knopf, 2002.

Andrade, Oswald de, 'Anthropophagist Manifesto'. *Latin American Literature and the Arts* no. 51: 65–8.

Andrews, Bruce, 'Electronic Poetics'. In *Cybertext Yearbook 2003*, edited by Markku Eskelinen, Raine Koskimaa, Loss Pequeño Glazier and John Cayley, 29–38: University of Jyväskylä, 2003.

Andrews, Jim, 'Arteroids'. www.vispo.com/arteroids/index.htm.

—'dbCinema'. vispo.com/dbcinema.

—'Games, Po, Art, Play, & Arteroids 2.03'. www.vispo.com/arteroids/onarteroids.htm.

—'Kandinsky 3: dbCinema Images Using the Flash Brush'. vispo.com/dbcinema/kandinsky3/intro/.

—'Stir Fry Texts'. collection.eliterature.org/1/works/andrews__stir_fry_texts/index.html.

—'Stir Frys and Cut Ups'. collection.eliterature.org/1/works/andrews__stir_fry_texts/text.html.

—email, Feb. 2009.

—email, May 2009.

—email, Sep. 2009.

—Vispo, vispo.com/dbcinema/huth/index.htm.

'Artist Biography: Angela Ferraiolo', www.brown.edu/Conference/Electronic_Literature_Organization/bios.php.

'Attention Span', en.wikipedia.org/wiki/Attention_span.

Bailey, R. W., *Computer Poems*: Potagannissing Press, 1973.

Baldwin, Sandy, 'Against Digital Poetics'. In *E-Poetry 2009: An International Digital Poetry Festival*. Barcelona, Spain, 2009.

—personal interview, Oct. 2008.

Bernstein, Charles, 'De Campos Thou Art Translated (Knot)'. epc.buffalo.
edu/authors/bernstein/essays/de-campos.html.

Bishop, Elizabeth (ed.), *An Anthology of Twentieth-Century Brazilian
Poetry.* Hanover, NH: Wesleyan UP, 1972.

Bootz, Philippe, 'Archipoenum'. code.google.com/p/archipoenum/.

—'Désir Insuffisant'. www.sitec.fr/users/akenatondocks/DOCKS-
datas_f/collect_f/auteurs_f/B_f/BOOTZ_F/TEXTES_f/Desir_Insuffi.
htm.

—'Discussion Comment E-Poetry 2009: An International Digital Poetry
Festival'. Barcelona, Spain, 2009.

—email, Jan. 2009.

Bopp, Raul, and Poty, *Cobra Norato E Outros Poemas.* Rio de Janeiro:
Civilização Brasileira, 1973.

Bouchardon, Serge, 'Presentation: Touch', www.utc.fr/~bouchard/
TOUCHER/docs/presentation-touch.pdf, 2009.

Bouchardon, Serge, Kevin Carpentier, and Stéphanie Spenlé, 'Touch'.
www.to-touch.com.

Bouchardon, Serge and Vincent Volckaert, 'Loss of Grasp'. lossofgrasp.
com/.

Breeze, Mary Anne (Mez), ' [Twitte]Reality_Fiction: 27/3/09'. version.
org/textuals/show/12.

—'_Feralc_'. netwurker.net/.

—'Dam(B)a({N}D{a})Ged Poetry'. delicious.com/TAGallery/EXHIBITION_
solo.wurk.

—'Id_Xor-Cism'. www.hotkey.net.au/~netwurker/xor/xor.html.

—'Twitter'. twitter.com/netwurker.

—'Video: Bruce Sterling's Keynote – at the Dawn of the Augmented
Reality Industry'. site.layar.com/company/blog/video-bruce-sterlings-
keynote-at-the-dawn-of-the-augmented-reality-industry/.

—email, Sept. 2010.

Brown, Nathan, 'The Function of Digital Poetry at the Present Time'.
newhorizons.eliterature.org/essay.php@id=11.html.

Burgess, Helen J., 'Introduction: Spaceworks'. www.hyperrhiz.net/
hyperrhiz03/6-introduction/20-introduction-spaceworks.

Cage, John, *I–VI.* Hanover, NH: Wesleyan UP, 1990.

Campos, Augusto de, email, Nov. 2006.

Cape, Douglas and Giles Perring, 'What We Will'. www.z360.com/what/.

Carpenter, Jim, 'Erika', etc.wharton.upenn.edu:8080/Etc3beta/Erika.jsp.

Cayley, John, 'P=R=O=G=R=a=M=M=a=T=O=L=O=G=Y: Wotclock'.
programmatology.shadoof.net/index.php?p=works/wotclock/
wotclock.html.

—'Program Notes for Imposition'. programmatology.shadoof.net/works/
impose/impositionProgNotes.pdf.

—'Wotclock' collection.eliterature.org/1/works/cayley__wotclock.html.

—personal interview, 2010

Cayley, John and Giles Perring, 'Imposition'. programmatology.shadoof. net/index.php?p=works/impose/impepoetry.html.

Cherry, Paul and Chris Morton, 'Googlism'. www.googlism.com/.

Coover, Robert, 'Afternoon Seminar'. Albany, NY, 1996.

Corrêa, Marina, 'Concrete Poetry as an International Movement Viewed by Augusto De Campos: An Interview'. www.scribd.com/ doc/36967268/Concrete-Poetry-as-an-International-Movement-Viewed-by-Augusto-de-Campos.

Douglas, J.Y. 'Understanding the Act of Reading: The Woe Beginner'. *Writing on the Edge* 2, no. 2 (1991): 112–25.

Drucker, J., 'The Virtual Codex from Page Space to E-Space'. In *A Companion to Digital Literary Studies*, edited by Susan Schreibman and Ray Siemens, 216–32. Malden, MA: Blackwell, 2007.

'Electronic Poetry Centre', epc.buffalo.edu/.

Emerson, Lori, 'It's Not That, It's Not That, It's Not That: Reading Digital Poetry'. In *E-Poetry 2003: An International Digital Poetry Festival*. West Virginia U, 2003.

Ferraiolo, Angela, 'Map of a Future War'. The New River, www.cddc. vt.edu/journals/newriver/08Fall/mapNewRiver/mapNewRiver/_ MapOfAFutureWar.html.

—'The End of Capitalism'. angelaferraiolo.com/end_of_capitalism/ capitalism.html.

—email, July 2010.

Fisher, Caitlin, 'Andromeda. Augmented Reality Poem'. Authoring Software, www.well.com/user/jmalloy/elit/elit_software.html#caitlin.

Flanagan, Mary, '[theHouse]'. collection.eliterature.org/1/works/ flanagan__thehouse.html.

Flusser, V., A. Ströhl and E. Eisel, *Writings*: University of Minnesota Press, 2002.

Friedman, Ken, 'Fluxlist and Silence Celebrate Dick Higgins'. www. fluxus.org/higgins/ken.htm.

Funkhouser, C., 'Dragon Fruit'. Poem from '13 States of Malaysia', 2006.

—'E-Poetry 2003: An International Festival of Digital Poetry'. (2003), www.wepress.org/E_Poetry_2003.pdf.

—'Encapsulating E-Poetry 2009: Some Views on Contemporary Digital Poetry'. dichtung-digital, www.brown.edu/Research/dichtung-digital/2009/Funkhouser/index.htm.

—*Prehistoric Digital Poetry: An Archaeology of Forms, 1959–1995*. University of Alabama Press, 2007.

Gervais, Bertrand, 'Is There a Text on This Screen? Reading in an Era of Hypertextuality'. In *A Companion to Digital Literary Studies*, edited

by Susan Schreibman and Ray Siemens, 183–202. Malden, MA:
Blackwell, 2007.

Ginsberg, Allen, *Cosmopolitan Greetings : Poems, 1986–1992*. New
York: HarperCollins, 1994.

Glazier, Loss Pequeño, *Digital Poetics: The Making of E-Poetries*.
University of Alabama Press, 2001.

Glenn, David, 'Divided Attention'. *The Chronicle of Higher
Education* (Feb. 28, 2010), chronicle.com/article/
Scholars-Turn-Their-Attention/63746/.

Guertin, Carolyn, 'Handholding, Remixing, and the Instant Replay:
New Narratives in a Postnarrative World'. In *A Companion to Digital
Literary Studies*, edited by Susan Schreibman and Ray Siemens,
231–49. Malden, MA: Blackwell, 2007.

Hayles, N. Katherine, *Electronic Literature: New Horizons for the Literary*.
Notre Dame, Ind.: University of Notre Dame Ward-Phillips, 2008.

Hennessey, Neil, 'Basho's Frogger'. collection.eliterature.org/2/works/
hennessy_frogger_jabber.html.

Hill, Crag, 'Crag Hill's Poetry Scorecard'. scorecard.typepad.com/.

'History of the World Wide Web', en.wikipedia.org/wiki/
History_of_the_world_wide_web.

Huth, Geof, 'Dbqp: Visualizing Poetics'. dbqp.blogspot.com/.

'Intermedia', en.wikipedia.org/wiki/Intermedia_%28hypertext%29.

'Internet Archive: Wayback Machine', www.archive.org/web/web.php.

Joyce, Michael, *Of Two Minds: Hypertext Pedagogy and Poetics*. Ann
Arbor: University of Michigan Press, 1995.

Keats, John, 'Letter to George and Tom Keats 27/12/1817'. In *Selected
Letters of John Keats*, edited by Grant F. Scott, 193. Cambridge:
Harvard U P, 2002.

Keep, Christopher, Tim McLaughlin and Robin Parmar, 'The Electronic
Labyrinth'. www2.iath.virginia.edu/elab/.

Kid, D. J.S. T. S. and P. Lunenfeld, *Rhythm Science*: MIT Press, 2004.

Kostelanetz, Richard and H. R. Brittain, *A Dictionary of the Avant-
Gardes*. New York: Schirmer Books, 2000.

Larsen, Deena, 'Carving in Possibilities'. collection.eliterature.org/1/
works/larsen__carving_in_possibilities.html.

—'Deena Larsen's Addicts Attic: Hypertext/New Media/Electronic Lit/
Possibilities'. www.deenalarsen.net.

—'The Pines at Walden Pond', www.cddc.vt.edu/host/deena/pines/.

—email, April 23 2009.

Lavagnino, John, 'Digital and Analog Texts'. In *A Companion to Digital
Literary Studies*, edited by S. Schreibman and R. G. Siemens,
402–18. Malden, MA: Blackwell, 2007.

Lehto, Leevi, 'Google Poem Generator'. www.leevilehto.net/google/
about.asp.

Malloy, Judy, Anna Couey and Douglas Cohen, 'A Conversation with
    Jim Rosenberg.' www.well.com/~couey/interactive/jim.html.
Manovich, Lev, 'Soft Cinema: Ambient Narrative'. www.softcinema.net.
McCauley, Carole Spearin, *Computers and Creativity*. New York:
    Praeger, 1974.
Memmott, Talan, 'Beyond Taxonomy'. In *New Media Poetics:
    Contexts, Technotexts, and Theories*, edited by Adalaide Kirby
    Morris and Thomas Swiss, 293–307. Cambridge, MA: The MIT P,
    2006.
—'E_Rupture://Codework.' 'Serration in Electronic Literature'. *American
    Book Review*,no. 22.6 (2001), talanmemmott.com/pdf/e_rupture.pdf.
Menezes, Philadelpho, 'Interactive Poems: Intersign Perspective for
    Experimental Poetry'. www.cyberpoem.com/text/menezes_en.html.
'Mezangelle', en.wikipedia.org/wiki/Mezangelle.
Mezei, Leslie, 'Computer Art …' *Artist and Computer* (1976), www.
    atariarchives.org/artist/sec7.php.
Montfort, Nick, 'Literary Games', nickm.com/writing/essays/literary_
    games.html.
Nelson, Jason, 'I made this. you play this. we are enemies'. www.
    secrettechnology.com/madethis/enemy6.html.
Nelson, Theodor H., *Geeks Bearing Gifts: How the Computer World Got
    This Way*. Sausalito, CA: Mindful Press : Distributed by Lulu.com,
    2008.
'Netartery', netartery.vispo.com/.
'Netpoetic', netpoetic.com/.
Niss, Millie, 'The Electronic Muse'. collection.eliterature.org/1/works/
    niss__oulipoems.html.
Object (Computer Science)', en.wikipedia.org/wiki/
    Object_(computer_science).
'Perl', en.wikipedia.org/wiki/Perl.
'Perlmonks', www.perlmonks.org/?node=Perl%20Poetry.
Perloff, Marjorie, *Radical Artifice: Writing Poetry in the Age of Media*.
    Chicago: University of Chicago Press, 1991.
—*Unoriginal Genius: Poetry by Other Means in the New Century*.
    Chicago: University of Chicago Press, 2010.
Perrin, Hélène and Arnaud Regnauld, 'Jim Rosenberg's Diagram
    Poems Series #3: A Few Preliminary Notes on Translation Issues'. In
    E-Poetry 2007: An International Digital Poetry Festival. Paris, France:
    Proceedings of E-POETRY 2007.
Pressman, Jessica, 'Navigating Electronic Literature'. newhorizons.
    eliterature.org/essay.php@id=14.html.
"Processing', processing.org/.
Rasula, J., 'From Corset to Podcast: The Question of Poetry Now'.
    American Literary History 21, no. 3 (2009): 660.

Rocha, João Cezar de Castro, 'Brazil as Exposition'. *Ciberletras*,no. 8
    (2002), www.lehman.cuny.edu/ciberletras/v08/rocha.html.
Rosenberg, Jim, 'Diagram Series 6: 6.4 and 6.10'. collection.eliterature.
    org/1/works/rosenberg__diagrams_6_4_and_10.html.
—'The Interactive Diagram Sentence: Hypertext as a Medium of
    Thought'. *Visible Language* 30.2, no. New Media Poetry: Poetic
    Innovation and New Technologies: 102–17.
—email, June 2008.
—email, Aug. 2008.
Salgado, Marcus, email, April 2007.
Sanford, Christy Sheffield, 'Flowerfall'. www.thing.net/~grist/l&d/
    bpnichol/tta/ch/coup.html.
Screen, Noelle, 'Re: Arteroids'. New Jersey Institute of Technology, 2010.
Seaman, Bill, 'Recombinant Poetics'. In *Media Poetry: An International
    Anthology*, edited by Eduardo Kac, 157–74. Chicago & London:
    Intellect Books, 2007.
Sérandour, Eric, 'Tue-Moi'. www.serandour.com/tue-moi.htm.
Sheperd, David, 'Finding and Evaluating the Code'. newhorizons.
    eliterature.org/essay.php@id=12.html.
Simanowski, Roberto, 'Understanding Text That Moves: Two Close
    Readings'. In *E-Poetry 2009: An International Festival of Digital
    Poetry*, 2009.
'Simulacra for Amiga', www.mobygames.com/game/simulcra.
Slattery, D. R. 'Alphaweb'. iat.ubalt.edu/guests/alphaweb/.
Sondheim, Alan, 'Blood.' www.alansondheim.org/Blood.txt.
—'On the Organization and Disorganization of Files'. www.
    alansondheim.org/00README1st.TXT.
—'Self Induction'. 2008.
—'Shudder'. 2010.
—personal interview, Brooklyn, NY, Feb. 2010.
Sondheim, Alan and Reiner Strasser, 'Dawn'. collection.eliterature.org/1/
    works/strasser_sondheim__dawn.html.
—'Tao'. collection.eliterature.org/1/works/strasser_sondheim__tao.html.
Sondheim, Alan and Sandy Baldwin, 'Second Life Performance'. Digital
    Poetry performance: E-Poetry 2009: An International Digital Poetry
    Festival. Barcelona, Spain, 2009.
Stefans, Brian Kim, 'the dreamlife of letters', collection.eliterature.org/1/
    works/stefans__the_dreamlife_of_letters.html.
—*Fashionable Noise: On Digital Poetics*. Berkeley: Atelos, 2003
—conversation with the author, May 2009.
Strehovec, Janez, 'The Poetics of Elevator Pitch'. In *E-Poetry 2009: An
    International Digital Poetry Festival*. Barcelona, Spain.
—'Text as Loop: On Visual and Kinetic Textuality', www.thefreelibrary.

com/Text+as+loop%3A+on+visual+and+kinetic+textuality.-
a0113683509.
Strickland, Stephanie, 'Quantum Poetics: Six Thoughts'. In *Media
Poetry: An International Anthology*, edited by Eduardo Kac, 25–44.
Chicago & London: Intellect Books, 2007.
—'Writing the Virtual: Eleven Dimensions of E-Poetry'. *Leonardo
Electronic Almanac* (2006), www.leoalmanac.org/journal/Vol_14/
lea_v14_n05-06/sstrickland.asp.
—Cynthia Lawson Jarmillo and Paul Ryan, 'Slippingglimpse.' www.
slippingglimpse.org/.
'The Electronic Literature Collection, Vol. 2'. collection.eliterature.org/2/.
'The Shannonizer', www.nightgarden.com/shannon.htm.
Tisselli, Eugenio, 'Synonymovie'. www.motorhueso.net/dcr/
synonymovie/synonymovie.htm.
—email, Feb. 2009.
'Ubuweb'. www.ubu.com/.
Vo, Jonathan, 'Re: Arteroids'. New Jersey Institute of Technology, 2010.
Wardrip-Fruin, Noah, 'Reading Digital Literature: Surface, Data,
Interaction, and Expressive Processing'. In *A Companion to Digital
Literary Studies*, edited by S. Schreibman and R. G. Siemens,
163–82. Malden, MA: Blackwell, 2007.
'Wayback Machine', en.wikipedia.org/wiki/Wayback_machine.
Wilks, Christine, 'Augmented E-Poetry at Elo_Ai'. *Netpoetic* (2010),
netpoetic.com/2010/06/augmented-e-poetry-at-elo_ai/.
Zakon, Robert Hobbes, 'Hobbes' Internet Timeline'. www.zakon.org/
robert/internet/timeline/.
Zellen, Jody, 'All the News That's Fit to Print'. www.
allthenewsthatsfittoprint.net/.
—'Without a Trace'. turbulence.org/Works/without_a_trace/.
—email, August 2009.
—'Digital Poetry Performance'. E-Poetry 2009: An International Digital
Poetry Festival. Barcelona, Spain, 2009.

Index

the things you forget. Susan Bensen, Zach Lorschaking you and you lie to him as you feel out JWTF on the victim in Mrs Maglans class, the idea that reading in a pub came to yourself after seeing that photo of him reaching for the beer, silhouetted by the window. Maurcad turns white faced and puts a pen in behind his eyes.

body poetics   CHORAL POETRY

ekphrasis as preservation   (random sets)

graphic design!   Samantha
                                              Gorman
        why don't we collaborate
                    w/ visual artists?
              youths, apple, sleekness
                   Ian Hatcher
trad of   history of CID in print
 Dada   sensitivity of aesthetics in digital age
 Fluxus   ↑
ALSO VISUAL ART   art world paying attention
  MOVEMENTS   to language again
                   ↑
NO PREREADING RISD Class (look at blogs)

print → more   Rob Wittig
 so text needs to be
   more legible to   "If you keep making things @ 1999,
   make up for   only people who were looking @ 1999
   loss of anticipatory will look at it now. + that is why
   reading   your community is insular."
              ↗

     also: losing an opportunity to make meaning
     (show work from class)
                        further cross-disciplinary learning
steve roggenbuck?   look @ ELO's chairmans
                            courses
 background on RISD)
  in footnote as
 academically induced ncd.
        JOANNA DRUCKER